Anonymous

Modified Specifications for triple Expansion twin-screw propelling Machinery

For U.S.S. San Francisco of 4,083 Tons Displacement and 19 Knots Speed

Anonymous

Modified Specifications for triple Expansion twin-screw propelling Machinery
For U.S.S. San Francisco of 4,083 Tons Displacement and 19 Knots Speed

ISBN/EAN: 9783337146153

Printed in Europe, USA, Canada, Australia, Japan

Cover: Foto ©berggeist007 / pixelio.de

More available books at **www.hansebooks.com**

MODIFIED SPECIFICATIONS

FOR

TRIPLE-EXPANSION

Twin-Screw Propelling Machinery

FOR

U. S. S. SAN FRANCISCO,

(CRUISER No. 5,)

OF

4,083 TONS DISPLACEMENT

AND

19 KNOTS SPEED.

WASHINGTON:
GOVERNMENT PRINTING OFFICE.
1888.

INDEX.

A

	Page.
Air-cocks, boiler	54
Air-ducts	47
Air-pressure gauges	58
Air-pumps	35
Ash-dumps	59
Ash-hoists	59
Ash-pans	47
Ash-pit doors	46
Ash-sprinklers	80
Attachments, auxiliary boiler	56
Attachments of valves to hull	71
Auxiliary boiler feed-pump	57
Auxiliary boiler attachments	56
Auxiliary boiler	55
Auxiliary engine stop-valves	70
Auxiliary exhaust-pipes	63
Auxiliary feed-tank	57
Auxiliary feed-pipes	64
Auxiliary feed-pump	59
Auxiliary steam-pipes	62

B

	Page.
Bars, lazy	47
Bearers and grate-bars	46
Bearings, reversing-shaft	21
Bearings, rock-shaft valve-motion	19
Bearings, stern-bracket	31
Bearings, stern-tube	30
Bearings, thrust	29
Bilge and double-bottom valve-boxes	71
Bilge and fire-pumps	60
Bilge injection-valve	39
Bilge-pumps, main	37
Bilge suction-pipes	66
Bleeder-pipes	63
Blowers	57

	Page.
Blowing-engines	58
Blow-pipes	64
Blow-valves, bottom	52
Blow-valves, surface	52
Boilers	41
Boilers and machinery, tests of	88
Boiler, auxiliary	55
Boiler air-cocks	54
Boiler attachments, main	50
Boiler bracing	42
Boiler clothing	55
Boiler drain-cocks	54
Boiler-heads	41
Boiler man-hole and hand-hole plates	44
Boiler material	41
Boiler protectors, zinc	54
Boiler pumping-out pipe	67
Boiler stop-valves	51
Boiler saddles	50
Boiler shells	41
Boiler tubes	42
Boiler tube-sheets	41
Bolts and nuts	81
Bottom blow-valves	52
Boxes, journal	80
Bracing, boiler	42
Brasses, crank-pin	14
Bridge-walls	46
Bulkheads and decks, pipes through	67
Bulkheads, shafts through	74

C

	Page.
Changes in plans	90
Check-valves, feed	52
Chests, valve	8
Circulating-plates	47
Circulating-pump connections	38

	Page.
Circulating-pumps	37
Circulating-pump engines	38
Clothing, boiler	55
Clothing, cylinder	8
Clothing, pipe	67
Coal-bunkers, pipes through	67
Cocks and valves	71
Cocks, gauge	53
Combustion-chambers	42
Condensers	32
Connections, circulating-pump	38
Connections, hose	64
Connections to valve rock-shafts	18
Connecting-rods	13
Couplings, shaft	27
Covers, cylinder	6
Covers, smoke-pipe	50
Covers, valve-chest	9
Crank-pin brasses	14
Crank-shaft pillow-block brasses	24
Crank-shaft pillow-blocks	23
Crank-shafts	25
Cross-head and slippers	13
Cross-head guides	23
Cross-heads, valve-stem	21
Cylinder covers	6
Cylinder casing, high pressure	3
Cylinder clothing	8
Cylinder drain-cocks	16
Cylinder linings	5
Cylinder man-hole covers	6
Cylinders, pump	81
Cylinder relief-valves	15

D

Decks and bulkheads, pipes through	67
Description, general	1
Desks, log	77
Distilling apparatus	60
Doors, ash-pit	46
Doors, furnace	45
Doors, uptake	48

	Page.
Drain-cocks, boiler	54
Drain-cocks, cylinder	16
Drain-pipes and traps	69
Drawings of completed machinery	90
Drawings, working	90
Dry-pipes	51
Ducts, air	47
Dumps, ash	59
Duplicate pieces	84

E

Eccentrics	17
Eccentric-rods	17
Eccentric-straps	17
Engine-frames	22
Engine-room instruments	76
Engine-room telegraphs	78
Engine-room water-service	65
Engines, securing in ship	85
Engine throttle-valve	14
Engines, blowing	58
Engines, circulating-pump	38
Engines, reversing	21
Engines, turning, and gear	69
Escape-pipes	64
Exhaust, main feed-pump	63
Exhaust-pipes	7
Exhaust-pipes, auxiliary	63

F

Feed check-valves	52
Feed-pump, auxiliary boiler	57
Feed-tanks	40
Feed-tanks, auxiliary	57
Feed-tank suction-pipes	66
Feed-suction from condensers	66
Fire and bilge-pumps	60
Fire-tool racks	58
Floor-plates	74
Frames, engine	22
Fronts, furnace	44
Furnaces	44
Furnace-doors	45
Furnace-fronts	44

G

	Page.
Gauges, air-pressure	58
Gauge-cocks	53
Gauges, steam, main boiler	53
Gear for working valves from deck	81
Gear, lifting	82
Gear, valve	16
General description	1
Grate-bars and bearers	46
Guides, cross-head	23

H

Hand-rails	75
Hand-hole and man-hole plates, boiler	44
Heads, boiler	41
High-pressure cylinder casing	3
Hose-connections	64
Hoists, ash	59
Hydrokineters	54

I

Indicator fittings and motions	77
Indicators, revolution	78
Injection-valve, bilge	39
Injection-valve, main	39
Instruments	83
Instruments, engine-room	76
Intermediate cylinder-casing	4

J

Jackets, steam	6
Journal-boxes	80
Joints, riveted	43

L

Labels on gear and instruments	72
Ladders	74
Lamps, supports for	81
Lazy-bars	47

	Page.
Levers, working, and gear	75
Lifting-gear	82
Link-blocks	18
Links	18
Links, suspension	18
Links, valve	20
Line-shaft	25
Linings, cylinder	5
Linings, valve-chest	9
Log-desks	77
Low-pressure cylinder-casings	4
Lubrication	79

M

Machinery and boilers, tests of	88
Machinery, drawings of, completed	90
Mandrels for white-metal bearings	80
Man-hole and hand-hole plates, boiler	44
Man-hole covers, cylinder	6
Main steam-pipes	61
Material and fitting of pipes	68
Materials and workmanship	86

N

Nozzles, stop-valve	44
Nuts and bolts	81

O

Office, superintending naval engineer's	89
Oil-drips	80
Oil-tanks	82
Omissions	91
Outboard-delivery valves, main	39

P

Painting	88
Pans, ash	47
Pillow-block brasses, crank-shaft	24
Pillow-blocks, crank-shaft	23

	Page.		Page.
Pillow-blocks, shaft	29	Record of weights	89
Pipe-clothing	67	Reversing-engines	21
Pipes, auxiliary-exhaust	63	Reversing-shaft bearings	21
Pipes, auxiliary-feed	64	Reversing-shafts	21
Pipes, auxiliary-steam	62	Revolution-indicators	78
Pipes, bilge-suction	66	Riveted joints	43
Pipes, bleeder	63	Rock-shafts, valve-motion	19
Pipes, blow	64	Rods, connecting	13
Pipes, dry	5	Rods, eccentric	17
Pipes, escape	64	Rods, piston	12
Pipes, exhaust	7		
Pipes, main-feed	64		
Pipes, main-steam	61	**S**	
Pipes, material and fitting of	68		
Pipes, sea-suction	65	Saddles, boiler	50
Pipes, smoke	49	Safety-valves	52
Pipes, suction, feed-tank	66	Salinometer-pots	54
Pipes through bulkheads and decks	67	Screw-propellers	31
		Sea-suction pipes	65
Pipes through coal-bunkers	67	Sea-valves	70
Pipes, thickness of	68	Securing engines in ship	85
Piston and slide-valves	10	Sentinel-valves	53
Piston-rods	12	Separators	67
Piston-rod stuffing-boxes	13	Shaft-couplings	27
Pistons	12	Shaft, line	25
Plans, changes in	90	Shaft pillow-blocks	29
Plates, circulating	47	Shafts	25
Plates, floor	74	Shafts, crank	25
Platforms, working	75	Shafts, propeller	26
Pots, salinometer	54	Shafts, reversing	21
Preliminary tests and trials	89	Shafts, rock, valve-motion	19
Propeller-shafts	26	Shafts through bulkheads	74
Propellers, screw	31	Shafts, thrust	26
Pump-cylinders	81	Shells, boiler	41
Pump, feed, auxiliary-boiler	57	Siren	73
Pump, relief-valves	70	Slippers and cross-head	13
Pumps, air	35	Smoke-pipe covers	50
Pumps, auxiliary-feed	59	Smoke-pipes	49
Pumps, circulating	37	Speaking-tubes	78
Pumps, fire and bilge	60	Sprinklers, ash	80
Pumps, main-bilge	37	Starting-valves	14
Pumps, main-feed	58	Steam-gauges, main boiler	53
		Steam jackets	6
R		Steam tube-cleaners	60
		Stems, valve	20
Racks, fire-tool	58	Stern-bracket bearings	31
Radiators	73	Stern-tube bearings	30
Rails, hand	75	Stern-tube stuffing-boxes	30

	Page.
Stop-valve nozzle	44
Stop-valves, auxiliary engine	70
Straps, eccentric	17
Stuffing-boxes	80
Stuffing-boxes, piston-rod	13
Stuffing-boxes, stern-tube	30
Stuffing-boxes, valve-stem	10
Suctions from double-bottom valve-boxes	66
Superintending naval engineer's office	89
Supports for lamps	81
Surface blow-valves	52
Suspension-links	18

T

Tank, auxiliary feed	57
Tank, feed	40
Tanks, oil	82
Telegraphs, engine-room	78
Tell-tales	79
Tests and trials, preliminary	89
Tests of boilers and machinery	88
Thrust-bearings	29
Thrust-shafts	26
Thickness of pipes	68
Throttle-valve, engine	14
Tools	83
Traps and drain-pipes	69
Trials and tests, preliminary	89
Tube-cleaners, steam	60
Tube-sheets, boiler	41
Tubes, boiler	42
Tubes, speaking	78
Turning-engines and gear	69

U

Uptakes	48
Uptake-doors	48

V

Valve, bilge-injection	39
Valve-chest covers	9
Valve-boxes, bilge and double-bottom	71

	Page.
Valve-chests	8
Valve-chest linings	9
Valve-gear	16
Valve-motion, rock-shaft bearings	19
Valve, main-injection	39
Valve-motion, rock-shafts	19
Valve-links	20
Valve, rock-shaft, connections to	18
Valve-stem stuffing-boxes	10
Valve-stems	20
Valve-stem cross-heads	21
Valves and cocks	71
Valves, attachments to hull	71
Valves, boiler-stop	51
Valves, bottom-blow	52
Valves, cylinder-relief	15
Valves, feed-check	52
Valves, gear for working from deck	81
Valves, main outboard-delivery	39
Valves, pump-relief	70
Valves, slide and piston	10
Valves, starting	14
Valves, sea	70
Valves, safety	52
Valves, sentinel	53
Valves, surface-blow	52
Ventilators	82

W

Walls, bridge	46
Water-gauges, main boiler	53
Water-service, engine-room	69
Weights, record of	85
Whistle	72
White-metal bearings, mandrels for	80
Working drawings	90
Working-levers and gear	75
Working-platforms	75
Workmanship and materials	86

Z

Zinc boiler-protectors	54

LIST OF PLANS ACCOMPANYING THESE SPECIFICATIONS.

A.—Arrangement of boilers, uptakes, &c.
B.—Longitudinal section of boiler, showing appliances for forced draught.
B¹.—End elevation and section of boiler.
C.—Auxiliary boiler.
D.—Air-pump and engine.
E.—Main bilge-pumps.
F.—H. P. cylinder.
G.—I. P. cylinder.
H.—L. P. cylinder.
I.—Alternative arrangement of L. P. valves.
J.—H. P. and I. P. pistons.
K.—L. P. piston.
L.—Piston-rod packing.
M.—Crank-shaft pillow-blocks and engine-frames.
M¹.—Crank-shaft pillow-blocks and engine-frames, original design.
N.—Piston-rod, cross-head and guides.
O.—Crank-shaft.
O¹.—Original design of crank-shaft.
P.—General plan of engines.
Q.—General arrangement of machinery.
R.—Connecting-rod.

SPECIFICATIONS

FOR

TWO DIRECT-ACTING,

TRIPLE-EXPANSION SCREW ENGINES AND BOILERS

FOR

U. S. S. SAN FRANCISCO.

REFERENCE BEING HAD TO THE ACCOMPANYING DRAWINGS, WHICH FORM A PART OF THESE SPECIFICATIONS.

GENERAL DESCRIPTION.

The propelling-engines will be placed in separate watertight compartments and will be duplicates—the high-pressure cylinder being the forward one in the forward engine and the after one in the after engine; the forward engine turning the port propeller. Each propelling-engine will have a high-pressure cylinder of 42 inches diameter, an intermediate cylinder of 60 inches diameter and a low-pressure cylinder of 94 inches diameter—the stroke of all pistons being 36 inches.

The barrels and heads of all cylinders are to be steam-jacketed.

The main valves will be of the piston type for the high-pressure and intermediate cylinders, and double-ported slide-valves for the low-pressure cylinder, all worked by Stephenson link-motion with double-bar links.

There will be two piston-valves for each high-pressure cylinder, two for each intermediate cylinder, and two double-ported slide-valves for each low-pressure cylinder.

Each piston will have a single piston-rod, with a cross-head working in slipper-guides.

The cylinder covers and the pistons for the high-pressure and intermediate cylinders will be of cast-steel. The low-pressure pistons will be of bronze. The cylinder liners will be of hard cast-iron.

The shafting, piston-rods, connecting-rods and moving parts generally are to be of forged open-hearth steel.

The ties forming the framing of engines and the pillow-blocks will be of forged steel. The crank-shafts will be made in interchangeable sections; all shafting will be hollow.

The condensers will be cylindrical, made entirely of sheet-copper and composition. Each condenser will have a cooling surface of about 7,276 square feet, measured on the outside of the tubes, the water passing through the tubes.

There will be for each propelling-engine a double horizontal double-acting air-pump worked by a vertical inverted-cylinder compound engine.

The circulating-pumps will be of the centrifugal type—one for each condenser—worked independently.

The main bilge-pumps will be operated by continuations of the air-pump piston-rods.

The propellers will be three-bladed, right and left, of manganese bronze.

There are to be four main boilers in two water-tight compartments, with four athwartship fire-rooms, the boilers being placed fore and aft. The boilers are to be of the double-ended horizontal return-fire tube type, 14 feet 8 inches in diameter, to carry a working pressure of 135 pounds per square inch. Each end of each boiler is to have three corrugated furnaces 3 feet 6 inches internal diameter. The total heating surface is to be about 19,480 square feet in main boilers, and a superheating surface of 258 square feet in the steam-pipes where they pass through the uptakes.

There will be one auxiliary boiler, with about 16 square feet of grate and 450 square feet of heating surface. There will be in each boiler compartment two feed-pumps, one for

the main and one for the auxiliary-feed system. Each system will be complete in itself, one entirely distinct from the other.

There will be two smoke-pipes, one for each compartment.

The forced-draught system will consist of blowers discharging into air-ducts under the fire-room floors, from which branch-ducts will lead to the ash-pits of furnaces. Means will be provided for closing the ash-pits when under forced draught and for preventing leakage of gases out of the furnace-doors. The air supply to each furnace when under forced draught will be regulated by a damper.

There will be a steam reversing-gear and steam turning-gear for each propelling-engine, steam ash-hoists, bilge and fire-pumps, a distilling apparatus, and such other auxiliary or supplementary machinery, tools, instruments or apparatus as may be described in the following specifications or shown in the official drawings.

HIGH-PRESSURE CYLINDER CASINGS.

The high-pressure cylinder casing of each main engine, including the steam and exhaust ports and passages, head, valve-chest and flanges, is to be made of the best cast-iron. It will have, in the cylindrical part, walls $1\frac{1}{8}$ inches thick. It will be faced and bored, as shown, for the reception of the working lining of cylinder and for the valve-chest lining.

The steam and exhaust ports will be smoothly cored to the dimensions shown in drawings.

The brackets at the inboard end, for attachment of engine-framing, will be well ribbed and truly faced to a plane perpendicular to the axis of the cylinder.

There will be a facing on the side of cylinder next the intermediate cylinder, properly bracketed and faced true with the line of cylinder; this will be firmly bolted to a similar facing on the intermediate cylinder.

There will be a bracket under the inboard end of cylinder, as shown in drawing, for securing the cylinder to engine-seating. The base of this bracket to be truly faced to a plane parallel to the axis of the cylinder.

There will be two supporting brackets under the outboard end of each cylinder, secured to engine-seating by holding-down bolts in slotted holes to allow for expansion.

The walls of the steam-passages will be properly stayed.

There will be facings, flanged and ribbed where necessary, for the attachment of the cylinder cover, valve-chest covers, throttle-valve, exhaust-pipe, starting-valve chest, starting-valve pipes, relief-valves, drain-cocks, indicator-pipes, jacket steam-pipes, jacket drain-pipes and oil-cups.

On the inboard cylinder-head there will be a bracket to support one end of cross-head slide; it will be truly faced parallel to the axis of the cylinder and to the face of the cylinder-seating.

INTERMEDIATE CYLINDER CASINGS.

The casing of each intermediate cylinder will be of the same material as that of the high-pressure cylinder, and similarly fitted, except in the following particulars:

There will be facings on each side, properly ribbed to correspond with those on the high-pressure cylinder on the one side and the low-pressure cylinder on the other side, to which they will be properly fitted and firmly bolted.

The walls of the cylindrical part of the casing will be $1\frac{3}{8}$ inches thick.

There will be facings for the engine-framing for the cross-head slide, cylinder cover, valve-chest covers, man-hole cover, for two high-pressure exhaust-pipes, intermediate exhaust-pipe, piston-rod stuffing-box, starting-valve chest, starting-valve pipes, relief-valves, drain-cocks, indicator-pipes, jacket steam-pipes, jacket drain-pipes, jacket safety-valve, receiver safety-valve, receiver live-steam pipe and oil-cups. There will be a 3-inch steam-pipe leading from main steam-pipe to the intermediate valve-chest, fitted with a 3-inch stop-valve.

LOW-PRESSURE CYLINDER CASINGS.

The casing of each low-pressure cylinder will be of the same material as that of the others. It will be fitted for two double-

ported balanced slide-valves, and will have a facing properly bracketed for bolting to the intermediate cylinder casing. The walls of the cylindrical part of the casing will be $1\frac{1}{2}$ inches thick. There will be facings for the diagonal braces with the rock-shaft bearings for the cross-head slide, cylinder-cover, valve-chest covers, man-hole covers, intermediate exhaust, low-pressure exhaust, piston-rod stuffing-box, starting-valve chest, starting-valve pipes, jacket drain-pipes, jacket safety-valve, jacket steam-pipe, relief-valves, drain-cocks, indicator-pipes, receiver safety-valve, receiver live-steam pipe and oil-cups.

CYLINDER LININGS.

The cylinder linings will be made of hard, close-grained cast-iron, of a satisfactory quality. They will be faced and turned to fit the cylinder casings. Each lining will have a bearing at about the middle of its length, as shown in drawings. Each lining will be secured at the outboard end by countersunk-headed through-bolts wherever possible, and elsewhere by countersunk-headed screws tapped into cylinder casing.

The joint at outboard end of steam-jacket to be made tight by calking the lining if necessary.

The inboard ends of linings to have inward flanges, and the joints to be made steam-tight, each by a thin copper ring secured to flange of lining and to cylinder casing by $\frac{5}{8}$-inch square-headed tap-bolts spaced about 3 inches between centers; these bolts to be set down on wrought-iron rings $1\frac{1}{4} \times \frac{1}{2}$ inch in cross-section, which reinforce the copper ring.

The linings, after having been secured in place in the casings, will be accurately bored to diameters of 42, 60 and 94 inches for the high-pressure, intermediate and low-pressure cylinders respectively, and to a thickness of 1 inch for the high and $1\frac{1}{8}$ inches for the intermediate and low-pressure cylinders.

The boring to be done with the cylinders in a horizontal position and while resting on their proper supports.

The linings will be counterbored at the inboard ends as shown in drawings.

CYLINDER COVERS.

The cylinder covers will be made of cast-steel, well ribbed on outside; the high-pressure and intermediate to be 1 inch thick and the low-pressure to be $1\frac{1}{4}$ inches thick. Each cover is to be of such form as to leave as little clearance as practicable, and will be recessed for the piston-shoe. It will be faced to fit the cylinder casing; also faced on outside of flange, bored and faced for the man-hole cover, and turned to receive the foundation of lagging. The covers will be secured to cylinder casings by $1\frac{1}{4}$-inch steel studs—there will be 34 for the high-pressure, 36 for the intermediate, and 44 for the low-pressure cover, equally spaced.

Holes will be drilled for jack-bolts and eye-bolts.

There will be a 15-inch man-hole in each high-pressure and intermediate, and a 25-inch hole in each low-pressure cover.

There will be a facing at lower part for cylinder drain-cock.

CYLINDER MAN-HOLE COVERS.

The man-holes in cylinder covers will have cast-steel covers of dished form over the 15-inch man-holes in covers of high-pressure and intermediate cylinder. They will be finished all over on the outside, faced and turned to fit the cylinder covers, and cored out for clearance of piston-rod nuts.

The covers of man-holes in low-pressure cylinder covers to be cored out for clearance of piston-rod nut, ribbed as shown in drawing, turned and faced to fit man-hole, and finished on outside of flanges.

All man-hole plates will be secured by 1-inch steel studs—10 for the high-pressure, 10 for the intermediate, and 14 for the low-pressure cylinder.

There will be holes drilled and tapped for jack-bolts and eye-bolts.

STEAM-JACKETS.

The cylinders will be steam-jacketed on the sides and heads. The space left around the working linings will form the

jackets, as shown in drawings. All ribs must be cored out so as to allow a free circulation of the jacket steam and a free drainage of the water of condensation. Steam for the jackets will be taken from the main steam-pipe in each engine-room on the boiler side of the throttle-valve.

A 1½-inch branch will connect the jacket of each high-pressure cylinder with main steam-pipe. A 2-inch branch, with a 2-inch adjustable spring reducing-valve adapted to pressures of from 50 to 135 pounds, will connect each intermediate jacket with main steam-pipe. A 2-inch branch, with an adjustable spring reducing-valve adapted to pressures of from 10 to 60 pounds, will connect each low-pressure jacket with main steam-pipe. Holes will be bored to connect the jacket spaces in heads with the jackets around the cylinder linings.

A drain-pipe will lead from the lowest part of each jacket to an approved automatic trap—one for each cylinder. Each trap to be provided with blow-through and bye-pass valves and discharge into the feed-tank, with a branch leading to bilge; both branches to have stop-valves.

Each drain-pipe will have a stop-valve near its jacket.

Each steam-pipe to jackets will have a separate stop-valve.

The drainage system of the jacket of each cylinder will be entirely independent as far as the trap discharge, from which point the drains may be in common. All pipes in the jacket-drain system will be put together by union joints, so as to be easily overhauled; joints to have projections to prevent packing choking the pipes.

There will be a 2-inch spring safety-valve on each intermediate and low-pressure jacket, capable of being adjusted from 60 to 135 pounds per square inch for the intermediate and from 10 to 60 pounds for the low-pressure jackets.

EXHAUST-PIPES.

The high-pressure cylinder will take steam between the valves, thus leaving the valve-stem stuffing-boxes and the

guides working in the exhaust pressure; there will, therefore, be two 17-inch copper exhaust-pipes leading from the ends of the high-pressure valve-chest to the ends of the intermediate valve-chest; there will be an exhaust connection 23 inches diameter between the intermediate valve-chest and that of the low-pressure cylinder; this connection will be of composition and fitted with a stuffing-box to provide for movement and to facilitate the making of the joint. A 30-inch copper pipe will connect the exhaust opening of the low-pressure cylinder with the proper nozzle on the condenser, with a 5-inch branch for bleeder-pipe. A light gridiron valve of composition will be fitted in main exhaust-pipes to shut the engine off from the condenser when the latter is used for auxiliary purposes, these valves to have their stems and operating gear detachable and stowed in convenient places.

CYLINDER CLOTHING.

The cylinders, valve-chests and exhaust-pipes, after being tested and finally secured in place in the vessel, are to be covered with approved incombustible non-conducting material and neatly lagged with black walnut or teak all over, with polished brass bands and brass lag-screws.

The lagging is to be made in removable sections over each cylinder cover, man-hole cover, valve-chest cover and starting-valve chest cover. The sections to be of such size as to be easily handled, and all parts marked. The lagging elsewhere is to be so secured as to be easily removed, replaced and repaired.

VALVE-CHESTS.

The valve-chests of the high and intermediate cylinders will each be fitted to receive two piston-valves. There will be openings at each end for inserting and removing the valves and working linings; the chests will be accurately bored and faced for the reception of the working linings. The valve-chest of the low-pressure cylinder will be fitted to receive two double-ported balanced slide-valves. There will be openings in the front or engine end of the valve-chest to withdraw or

place the valves on their seats; the faces will be accurately faced to a true surface and made perfectly steam-tight before the insertion of the linings. All steam and exhaust passages must be thoroughly cleaned out, and care must be taken that the passages are nowhere contracted to less than the specified areas. Each intermediate and low-pressure valve-chest will have a 3-inch adjustable spring safety-valve of an approved pattern.

VALVE-CHEST LININGS.

There will be working linings to each of the piston-valves. They will be made of hard, close-grained cast-iron of a satisfactory quality. They will be accurately turned and faced to fit the castings, and will be accurately bored to an internal diameter of 12 inches for each high-pressure, and 20 inches for each intermediate valve, leaving the walls about 1 inch thick, as shown in drawings. They will be forced into place, making all joints perfectly tight, and will be secured in place by screws tapped half into the linings and half into the casings. The steam-ports will have alternating right and left diagonal bridges, about 1 inch wide, leaving a clear port opening of 100 square inches in each high-pressure valve-chest lining and 210 square inches in each intermediate valve-chest lining. There will be a working lining for each valve in the low-pressure valve-chest, made of hard, close-grained cast-iron, properly fitted to the face on the casing, secured by steel tap-bolts, with slotted heads, counter-bored so as to finish $\frac{1}{4}$ inch below the working surface. All parts will have their edges finished accurately to given dimensions. The combined area of steam-port through the low-pressure valve-seats will be 700 square inches.

VALVE-CHEST COVERS.

The valve-chest covers will be made of cast-steel, in dished form, ribbed, as shown in drawings.

The flanges will be turned to fit the openings in valve-chests and finished on outside. Each inboard cover will have a stuffing-box cast in, and will have a composition

bushing, well secured, at bottom of stuffing-box; this bushing to be grooved circumferentially to serve as a wearing surface for valve-stem.

Sufficient metal will be left in the valve-chest cover at the side of the stuffing-box for a $\frac{1}{2}$-inch oil passage; this passage will be drilled from the outside of the cover, and will communicate with the inner circumferential groove of the bushing.

Each outboard cover will be bored to receive a composition bushing, as shown in drawings, which will be grooved in same manner as bushing at inboard end, and provided with a similar oil passage leading through cover to the inner circumferential groove.

The outboard valve-chest covers of the intermediate and low-pressure valve-chest will each have an outside sleeve bolted to the cover, there being a facing on the cover to secure this sleeve.

The outboard cover of the high-pressure cylinder will have the sleeve cast on.

Valve-chest covers will be secured by $1\frac{1}{8}$-inch bolts, spaced not over 6 inches, with finished wrought-iron nuts.

VALVE-STEM STUFFING-BOXES.

The valve-stem stuffing-boxes will be cast in the inboard covers of valve-chests, and will be designed and bored out to the proper size to receive Watson's metallic packing. The packing-rings and shell will be made of bronze. The bottoms of the stuffing-boxes will be formed by the bushings before specified.

PISTON AND SLIDE-VALVES.

Piston-valves will be made of composition for the intermediate and of cast-iron or cast-steel for the high-pressure cylinders; each valve will be in two parts, each of these parts consisting of a piston with follower, wearing-ring and two packing-rings. The slide-valves for low-pressure cylinder will be made of close-grained, strong cast-iron, of not less

than 20,000 lbs. tensile strength, cast with the proper ports and sleeve for the valve-stem. There will be a facing on the upper side to receive the working face of the balance-ring, the inside area of which will equal the exhaust opening. This ring will be made of U-sections, as shown on drawing, of copper, with a composition wearing-face. The joint of the cover that carries the balance-ring will be a ground joint, made perfectly tight, without any soft packing.

In the piston-valves the followers will be of the same material as the pistons, and secured in place by steel bolts with wrought-iron nuts and brass split-pins. The follower-bolts will pass through lugs on inside of shell of each intermediate and low-pressure valve. The heads will be so formed and fitted as to prevent turning.

The follower-bolts of high-pressure valves will be T-headed, with heads pocketed in shells of valves.

The wearing-rings will be made of close-grained cast-iron; they will be finished to a neat end fit between the valve-piston and follower, but will be a loose side fit. They will be smoothly and accurately turned and faced for the reception of the packing-rings. The packing-rings will be of best cast-iron, cast larger than the bore of valve-chest lining, cut obliquely, sprung in and turned to proper diameter, tongued and fitted in place.

The packing-rings will have wearing-faces $1\frac{1}{4}$ inches wide, and will be $\frac{3}{4}$ inch thick.

The two ends of each valve will be separated, when in place on their valve-stem, by a cast-steel distance piece, which will be of such length as to make the steam and exhaust laps as follows:

	High-pressure.		Intermediate.		Low-pressure.	
	Inboard.	*Outboard.*	*Inboard.*	*Outboard.*	*Inboard.*	*Outboard.*
Steam-lap	$2\frac{1}{4}$	$2\frac{11}{16}$	$2\frac{1}{4}$	$2\frac{11}{16}$		
Exhaust-lap	$+\frac{5}{16}$	0	$+\frac{5}{16}$	0		
Steam-lead	$+\frac{3}{16}$	$1\frac{1}{4}$	$+\frac{3}{16}$	$1\frac{1}{8}$		

PISTONS.

The high-pressure and intermediate pistons will be made of cast-steel, with single shells. The low-pressure pistons will be made of composition, with double shells, well ribbed.

The followers are to be of cast-steel, each made in two parts, the lower part forming a shoe and wearing-surface for pistons. They will be secured by $1\frac{1}{4}$-inch bolts, 10 for each high-pressure, 14 for each intermediate, and 20 for each low-pressure piston.

The follower-bolts will be steel studs screwed into the pistons; the bodies of the studs will be square, passing through square holes in the followers. The follower-bolt nuts will be of wrought-iron, finished and case-hardened; each nut will be secured in place by a brass split-pin of ample size.

Each piston will have two segmental packing-rings of hard cast-iron, each $\frac{5}{8}$ inch wide and $\frac{3}{4}$ inch thick, the sections to be loosely doweled together.

The packing-rings will be set out by steel springs of approved pattern. The springs will be of the best spring-steel, properly tempered. The pistons must work steam-tight without undue friction.

The wearing-shoe of each piston, forming the lower part of the follower, will extend for about one-third of the circumference at bottom of piston.

The holes for the bolts in the lower part of each follower will be slotted vertically to allow of adjustment of the shoe. The bottom of each shoe will be lined with strips of approved anti-friction metal. The lengths of bearing of the shoes, in the direction of the axes of the cylinders, will be 7, 9 and 11 inches for the high, intermediate and low-pressure cylinders respectively. The cores of low-pressure pistons must be thoroughly cleaned out, core-plugs screwed in and locked, and the pistons tested for tightness.

PISTON-RODS.

The piston-rods are to be of forged steel $7\frac{1}{4}$ inches diameter;

they will be turned to fit the piston with collar, as shown, and fitted each with a steel nut secured by a screwed stop-pin. The parallel parts are to be smoothly and accurately turned. The inboard end of each piston-rod will be finished to fit the cross-head with a collar and steel-nut fitted and secured same as in piston end. The collars on each end of piston-rods will be 9 inches diameter and 2 inches thick, well filleted, and recessed in the pistons and cross-heads.

CROSS-HEAD AND SLIPPERS.

The cross-heads will be of forged steel, as shown on drawing. It will be bored to receive the piston-rod end and have two journals, each $8\frac{1}{2}$ inches in diameter and 8 inches long; a conical hole will be bored in each journal end to reduce the weight; a composition or bronze slipper-block will be fitted and bolted under the cross-head as shown. A composition guide-gib in two parts will be fitted to the bottom of each slipper, flanged over the slipper at each end and held up by locked tap-bolts. Each gib to be finished with white-metal slabs, dovetailed, set in place by 25 tons to the square inch hydraulic pressure, and provided with oil-grooves.

PISTON-ROD STUFFING-BOXES.

The piston and stuffing-boxes will be all alike, the shells being of composition and made in halves to provide for slipping over the collars on the piston-rods. They are to be fitted with Watson's metallic packing, arranged as shown on full-size drawing.

CONNECTING-RODS.

The connecting-rods will be made of forged steel and, with their fittings, finished all over. Each rod will be forked at the cross-head end, and fitted with cross-head brasses in accordance with the drawings; the crank end will be finished with steel cap, and bolts with recessed nuts and set-screws. The rod and cap will be separated by chipping-pieces and brass liners, held

in place by dowels. The caps and brasses at each end will be fitted with eye-bolts for handling. The connecting-rods will be 6 feet between centers, 6 inches diameter at cross-head end, and 7 inches at crank-pin end.

CRANK-PIN BRASSES.

The crank-pin brasses are to fit the connecting-rod ends and caps neatly. They will have proper clearance at top and bottom, and will be lined with approved anti-friction metal. Each brass will be held in place by four tap-bolts, two at top and two at bottom, with set-screws.

The brasses are to be $\frac{1}{8}$ inch shorter than the pins.

ENGINE THROTTLE-VALVE.

There will be on each engine a gridiron slide-valve for a throttle, the casing of the valve to be bolted to the high-pressure valve-chest. There will be four openings through valve-seat and three through valve; each opening will be about 2 x 18 inches in the clear; the valve, when closed, to lap over the seat on each side about $\frac{1}{4}$ inch; the ports in valve and seat to have their edges beveled back from the faces.

The valve, stem and casing to be made of composition; the casing to be made in two parts, and joined by flanges near the line of valve-face; the shell of casing to be $\frac{1}{2}$ inch and flanges $\frac{3}{4}$ inch thick.

The valve to be stiffened by two ribs on its back; to have a stem passing through a stuffing-box and operated through suitable gear by a composition hand-wheel 18 inches in diameter with a winch-handle at the working-platform.

The thread on the valve-stem will have a pitch of 2 inches. An index to be fitted to show position of valve.

STARTING-VALVES.

There will be a starting-valve for each cylinder of the propelling-engines. Each valve is to be complete in itself, with both steam and exhaust-ports. The valves, chests and covers

will be of composition. Each valve will have a rib on its back, working close under the cover of its chest, to prevent the valve rising off its seat; the valves to be securely fastened to their stems; the valve-chests to be bolted to the facings provided for them. Steam for the starting-valves will be taken from the main steam-pipes outside the throttle-valves by a pipe having a branch to each valve. There will be a stop-valve in this pipe, to be worked from the working-platform; also a stop-valve close to each valve-chest. Each starting-valve will connect with each end of its cylinder by a 2-inch copper pipe. These valves will all exhaust into a 2-inch pipe leading to the condenser, the branch from each valve having a stop-valve close to the valve-chest. Each steam-port of each starting-valve will have an area of about 3 square inches. Each starting-valve will be worked by a lever at the working-platform; these levers to be placed in the same order as their respective valves, and to move in the same direction as the desired motion of the pistons. The valves are to be in middle position when their levers are vertical. The starting-valve faces and seats to be scraped to perfect steam-tight joints.

CYLINDER RELIEF-VALVES.

There will be an adjustable spring relief-valve on each end of each main cylinder of the following diameters: High-pressure relief-valves, $3\frac{1}{2}$ inches; intermediate relief-valves, $3\frac{1}{2}$ inches; low-pressure relief-valves, 5 inches. Pipes will lead from the relief-valves to the bilge with easily-broken joints.

These valves to have nickel seats or their equivalent, and the valve-fittings to be so constructed that the valves can be easily overhauled without slacking the springs, and so that steam will not come into contact with the springs. The springs will have approved means of adjustment, and will be long enough to allow the valves to open to their full extent without unduly increasing the load. The valves to be guided

by loosely-fitting wings. The springs are to bear on shoulders on spindles which fit loosely in sockets recessed in the backs of the valves. These spindles to be so fitted that the valves can be moved by the application of a lever. The valves to be fitted with casings, which will prevent danger of people being scalded by hot water from the cylinders. Suitable fulcrums to be on casings for the application of levers for working the valves; one lever to be furnished for each engine-room. All springs to pass a satisfactory test.

The spring-casing of each valve to be fitted with a suitable lock; all locks to have similar keys.

CYLINDER DRAIN-COCKS.

There will be a drain-cock on each end of each main cylinder of the following diameters of opening: High-pressure drain-cocks, $1\frac{1}{2}$ inches; intermediate drain-cocks, $1\frac{1}{2}$ inches; low-pressure drain-cocks, 2 inches. The cocks to be perfectly tight without undue friction. The drain-cocks of the high-pressure and intermediate cylinders of each engine are to be worked by a single lever, and those of the low-pressure cylinder by a separate lever, both at the working-platform. All the drain-cocks of each engine will discharge into a pipe leading to the fresh-water side of the condenser, with a branch to the bilge. The cock connecting the two branches must be such that one branch must always be open, but without permitting air to enter the condenser. Small drain-cocks will be fitted to the lowest parts of drain-pipes.

The drain-cock at the inboard end of each cylinder will be attached to a brass pipe which will pass loosely through the cylinder-casing and be tapped into the cylinder-lining. This pipe will be fitted with a brass nut and copper washer set up on the cylinder-casing so as to make a tight joint, but allow of the expansion of the cylinder-liner.

VALVE-GEAR.

The valve-gear is to be of the Stephenson type, and must

be so adjusted as to cut off steam at such a point as shall give the maximum engine power.

The links to be of the double-bar type, operating the valves through the equalizing-bars and valve-motion rock-shafts.

The links to be down in forward gear.

ECCENTRICS.

The eccentrics are to be of cast-steel, each in two parts, securely bolted together. They will be bored to fit the shafts snugly and will be held in place each by a key and set-screw. They will be all alike and truly finished to dimensions as follows: $5\frac{1}{4}$ inches eccentricity and 29 inches greatest diameter; they will each be 5 inches wide, and will have their faces recessed $\frac{1}{2}$ inch wide and $\frac{1}{8}$ inch deep on each side.

ECCENTRIC-STRAPS.

The eccentric-straps will be made of composition, truly bored and faced to fit eccentrics. The two parts will be held together by two 2-inch bolts; these bolts to have square heads and hexagonal wrought-iron nuts, locked in place; the bolts to pass through lugs on straps. Channeled brass chipping-pieces will be fitted between the two parts of each strap, notched over the bolts, and held in place by dowels or their equivalents. The back part of each strap will have a facing for the foot of the T-head of the eccentric-rod. The strap will be stiffened by a rib 3 inches deep and $1\frac{3}{4}$ inches wide in the line of eccentric-rod, decreasing in depth towards the sides.

There will be two 2-inch stud-bolts screwed into the back of each strap to secure the T-head of the eccentric-rod; these bolts to have finished wrought-iron nuts.

ECCENTRIC-RODS.

There will be one wrought-iron eccentric-rod for each strap, with a T-head at one end to secure it to its strap; the link end to be forked and fitted with brasses, strap, gib and key, or caps and bolts, as may be found most suitable. They shall

all be alike, except as to being right and left for off-set; they shall be $2\frac{1}{2}$ x 5 inches at the strap end and $2\frac{1}{2}$ x $3\frac{1}{4}$ inches at the link end.

LINKS.

The links will be of the double-bar pattern, made of forged steel, finished all over. All links will have the same mean radius and a cross-section of 2 x 5 inches. The links will be down when in forward gear. The link-pins to be forged in the links and finished to $3\frac{1}{2}$ inches diameter and 3 inches long; the pin at upper end of each to be extended to form the suspension-pin, which will be $2\frac{1}{2}$ x $2\frac{1}{2}$ inches. Each pair of bars will be secured by a shouldered steel stud at each end, and fitted with a finished wrought-iron nut at each end; the bars will be fitted $6\frac{1}{2}$ inches face to face.

LINK-BLOCKS.

The link-blocks will be made of forged steel, fitted with composition gibs, arranged to receive liner for taking up the wear, the gibs to fit accurately on curved sides of the links. The middle of each link-block to be turned to form a 4 x 4-inch link-block pin, terminating at each end in a pair of jaws $8\frac{1}{2}$ inches wide, to span corresponding bar of link.

SUSPENSION-LINKS.

Each Stephenson link will be suspended from the correponding arm of the reversing-shaft by two flat-sided, forged-steel suspension-links; to be $2\frac{1}{2}$ x $3\frac{1}{2}$ inches at the middle and $2\frac{1}{2}$ inches diameter at ends.

The ends of the links to be fitted with composition bushings, bored to fit suspension-pins on main links and pins on reversing-shaft arms.

CONNECTIONS TO VALVE ROCK-SHAFTS.

The motion of each link-block will be transmitted to the valve rock-shaft by a connecting-link, one end of which will

be so guided by a radius-link as to minimize the slip of the link-block when in forward gear. These connecting-links will be made of cast-steel, in form as shown, finished all over; each link will have adjustable brasses at link-block and rock-shaft connections, and a composition bushing at radius-link connection.

Each radius-link will have a bushed eye at upper end working on a fixed pin, and a double eye and pin at lower end to fit the bushing in the connecting-link. The pins for the fixed centers of these links will be carried by the rock-shaft bearings. Dimensions and positions of centers to be as shown on diagram of valve-gear.

VALVE-MOTION ROCK-SHAFTS.

The valve-motion rock-shafts and all the arms will be of forged steel and finished all over. Each will be carried in two bearings as shown; each will have three arms, two to connect with valve-stems and the other to connect with link-motion, which will have a pin forged on to finish for a journal 4 x 4 inches. The arms connecting with the valve-stems will each have a jaw to span the valve-stems, each side having a pin turned to $2\frac{1}{4}$ x $2\frac{1}{4}$ inches. The rock-shafts will be finished with raised portions to receive the arms and form collars for the bearings. The rock-shaft for low-pressure valves will be finished to a diameter of 10 inches in the arms and 8 inches in the journals and body, and will have a 4-inch hole bored through it. The rock-shafts for the high-pressure and intermediate valves will be finished to a diameter of 8 inches in the arms and 6 inches in the journals and body, and will have a 3-inch hole bored through them. The arms will be pressed on with a pressure of 10 tons per inch of diameter, and secured by one key.

VALVE-MOTION ROCK-SHAFT BEARINGS.

The valve-motion rock-shaft bearings will be of forged iron and finished all over; they will rest upon and be bolted

to the square portion of the upper struts of the engine-framing, as shown in drawings; they will be fitted with adjustable brasses and caps. Those for the low-pressure valve-motion rock-shafts will be forged with and form part of a pair of diagonal ties, extending from the main pillow-blocks to the front-head of low-pressure cylinder, as shown in the drawings.

VALVE-LINKS.

The valve-links, two for each valve, will be of forged steel, finished all over, 10 inches between centers.

The end connecting to valve-stem will be fitted with a composition bushing; the end connecting with rock-shaft arm will have adjustable brasses. Each valve-link will carry a wiper oil-cup at each end that will take oil from a stationary oil-cup, properly secured and provided with suitable wicks.

VALVE-STEMS.

The valve-stems will be of forged steel; 4 inches in diameter at the stuffing-box and reduced to $2\frac{3}{4}$ inches where they pass through the valves. There will be a hole bored in the valve-stems at their inboard ends, $2\frac{5}{8}$ inches diameter. The stems to be tapped at their inboard ends, each with a steel stud screwed into it and pinned in place, the studs to be 3 inches diameter. The studs will be turned down, where they pass through the valve-stem cross-head, to $2\frac{1}{4}$ inches diameter, and fitted with feathers to prevent valve-stem cross-head turning on stem. The ends of the studs will be threaded and fitted with finished wrought-iron nuts.

A composition washer $\frac{1}{2}$ inch thick to be placed between each cross-head and shoulder of valve-stem for valve adjustment. There will be at the outboard end of each valve-stem a steel distance-piece and composition nut to secure the valve in place; this nut to fit a right-handed thread on valve-stem and to be locked by a nut on a left-handed thread on valve-stem.

VALVE-STEM CROSS-HEADS.

The valve-stem cross-heads will be of cast-steel, finished all over; to have a pin at each end $2\frac{1}{4} \times 2\frac{1}{4}$ inches. The boss of each cross-head to be bored to fit a stud on valve-stem and to have a key-way to fit the feather on same.

REVERSING-SHAFTS.

There will be a forged-steel reversing-shaft in three sections to facilitate removal, bolted together by solid forged couplings. The arms for operating the links and the arm attached to the reversing-gear will be of forged steel; the arms for the high-pressure link and the low-pressure link will be keyed on the ends of the shaft, and the arm for the intermediate link will be keyed on the coupling; the arm for reversing-gear will also be on the coupling. The reversing-arm will be arranged to receive a sliding block, and will have a cap for disconnecting; this arm will be designed so as to span the guide and cross-head of the reversing-engine. The reversing-shafts will be 6 inches in diameter in the journals and body and 7 inches in the arms, and will have a 3-inch hole through its whole length; each link-arm will be provided with a screw-adjustment and slot for the suspension-pin, so that any link can have its position changed independently of the others, the slot to be set at such an angle as to bring all the links in full gear when reversed, independent of the position of any of the pins in the slot, and to be so arranged that it can be adjusted for any link from the working-platform by means of a socket-rod.

REVERSING-SHAFT BEARINGS.

The reversing-shaft bearings will be made with their caps of composition, the bearings to be securely bolted to the square part of the lower struts of the engine framing.

REVERSING-ENGINES.

Each reversing-engine will consist of a steam-cylinder with a hydraulic controlling cylinder. The steam-cylinder will be

20 inches diameter, with a stroke of about 27 inches, and the controlling cylinder 9 inches diameter; it will occupy a recess between the high-pressure cylinder and the intermediate cylinder, on facings provided in the web-casting between the cylinders. The steam piston-rod will be secured to a steel cross-head at lower end, working in suitable guides, and connecting by sliding-blocks with the arm on the reversing-shaft; the piston-rod on the upper end will pass into the controlling cylinder and continue through it, having the same diameter at each end; the controlling cylinder will be of composition. The valve of the steam-cylinder will be of the piston pattern of composition, working in a composition-lined valve-chest; there will be a bye-pass valve on the hydraulic-cylinder, which will be worked by a continuation of the stem of the steam piston-valve; these valves will be worked by a system of differential levers, the primary motion being derived from the hand-lever on the working-platform, and the secondary motion from a pin on the reversing-arm, all parts to be so adjusted that the reversing-engine shall follow the motion of the hand-lever and be firmly held when stopped; a suitable clamping device, easily worked from the platform, to be fitted so that the position of the links shall not be affected by any leakage of the hydraulic-gear. There will be a stop-cock in the bye-pass pipe of the hydraulic-cylinder; the pump for reversing by hand will be fitted to the side of the hydraulic-cylinder, with its lever convenient to the working-platform; the bye-pass pipes will pass through the valve-box of the hand-gear in such a way as to leave the hand-arrangement in gear all the time.

ENGINE-FRAMES.

The framing of the engines will consist of forged-steel or wrought-iron struts, two of which will connect each crank-shaft pillow-block to the corresponding cylinder. The body of the struts will be 6 inches diameter, the portion that receives the bearings for working-shaft and reversing-shaft will

be 7 inches square; each strut will have a flange at the cylinder end $2\frac{1}{4}$ inches thick and 15 inches square, faced on both sides, and bolted to the cylinder casing by 8 bolts $1\frac{1}{2}$ inches diameter, bolt-holes to be slotted to permit of a slight vertical adjustment; abutments will be cast in the cylinder faces to receive keys, top and bottom. The crank-shaft end of each strut will be enlarged to 8 inches for a distance of 12 inches, with a collar at the end of 11 inches diameter and 2 inches thick; the end will be bored out to receive the end of the pillow-block and binder-bolt. The pillow-blocks to be so made that the back brasses can be adjusted without the addition or reduction of liners, according to details hereafter worked out and approved by the Navy Department.

CROSS-HEAD GUIDES.

The lower portion of cross-head guides will be made of close-grained cast-iron, with circulating water passages cast through it, as shown on drawing; one end will be secured to a bracket cast for that purpose on the cylinder casing; the other end will rest on a facing cast on the steel foundation plate, as shown. The wearing-surface will be 23 inches wide and 4 feet 8 inches long; there will be a steel backing-guide, bolted on each side of each cross-head guide, the joint to be on the line of the face so that the slipper-gibs can be withdrawn sideways, or may be lined up without being removed. There will be 7 reamed bolts $1\frac{1}{4}$ inches diameter for each backing-guide; the guides will be well finished and secured in proper alignment; oil-grooves and proper oil-boxes to be fitted.

CRANK-SHAFT PILLOW-BLOCKS.

The crank-shaft pillow-blocks, two for each section of crank-shafts, will be made of forged steel or wrought-iron; the base will be faced on the bottom, and the block will be tool-finished all over; there will be a groove cut in the bottom perpendicular to the line of the shaft, which will fit over a tonny cast on the upper face of the foundation plate. There will also be

side seats for taper liners or keys, for adjustment, as shown. The holding-down bolts, six in number, will pass through slots in the taper liners; the taper will be on the under side of the liners to admit of a horizontal adjustment of the pillow-block without any vertical movement. Each pillow-block will have a forged-steel cap which, together with the pillow-block, will be bored and faced for the brasses, and will be accurately fitted to the jaws of pillow-block; the cap-bolts will be of forged steel $4\frac{1}{2}$ inches diameter in the thread, 5 inches in the struts, 6 inches in the collars, and $3\frac{3}{4}$ inches elsewhere, as shown in drawing. The outer end will be threaded and fitted with finished wrought-iron nuts, with collars and set-screws. All the necessary oil-holes and holes for water service to be provided. The pillow-blocks for low-pressure engine to be shaped on top to receive the foot of a diagonal brace, as shown.

CRANK-SHAFT PILLOW-BLOCK BRASSES.

The crank-shaft pillow-block brasses will be made of composition, lined with approved anti-friction metal. Each pair of brasses will be bored for a crank-shaft journal, and turned to fit the pillow-block and cap; they must be easily removable when the shaft is in place.

Each brass will be held in place by four $1\frac{1}{4}$-inch tap-bolts passing through pillow-block or cap, as the case may be, and screwed into the brass, two at top and two at bottom. The brasses will be separated by channeled brass chipping-pieces, each secured by two lugs and tap-bolts, and easily removable when the brasses are in place. There will be oil-grooves about $\frac{1}{2}$ inch wide and about $\frac{1}{4}$ inch deep between the slabs of anti-friction metal, extending to within $\frac{1}{2}$ inch of each end of the brasses, where the brass will form a stop to prevent the oil running out.

The brasses will be finished all over and parted, as shown in drawing. Each brass will be tapped and fitted with eye-bolts for handling. The back brass of each pillow-block will

be tapped, as shown in drawing, for a 1½-inch composition pipe, the pipe to be screwed into the brass, and to extend through the back of pillow-block and to be connected with the engine-room water-service pipes. A suitable valve will be placed in this pipe in an accessible position.

Each front brass will have a hand-hole to correspond to the hand-hole in the pillow-block cap.

SHAFTS.

All the crank, line, thrust and propeller-shafts are to be of steel. Each length will be forged solid in one piece, and will have a hole drilled axially through it from end to end.

All shafts are to be finished all over.

CRANK-SHAFTS.

There will be three sections of crank-shafts for each propelling-engine, all alike and interchangeable. Each section will have a crank of 18 inches throw and will have a coupling at each end. The length of each section of shaft will be 8 feet 4 inches over all. There will be two journals, one on each side of the crank, each 14¼ inches in diameter.

The shaft will be increased to 15 inches diameter at the eccentric seatings. The crank-pins will be 15 inches diameter and 16 inches long. The crank-webs will each be 17 inches wide and 10½ inches thick, the webs to be beveled, as shown in drawing. The crank-pins must be accurately parallel to the main journals. All journals are to be smoothly and accurately turned, and when finished will be tested and their accuracy proved. There will be a hole 7 inches in diameter bored axially through each shaft and crank-pin. When bolted together the cranks will be at angles of 120° to each other, the intermediate to follow the high-pressure and the low-pressure to follow the intermediate.

LINE-SHAFT.

The port crank-shaft will be connected to the thrust-shaft by a line-shaft in two lengths. It will be 13¼ inches diameter,

with a 7-inch axial hole. It will rest in a pillow-block in the forward engine compartment, one in after engine compartment, and one in shaft-alley. It will also have a spring-bearing in after engine compartment between cross-head slides of air-pump.

The starboard crank-shaft will be connected to the thrust-shaft by a line-shaft in one length, with one pillow-block in shaft-alley. There will be a distance of $\frac{3}{8}$ inch between line-shafts and thrust-shafts.

THRUST-SHAFT.

Each thrust-shaft will be $13\frac{1}{4}$ inches diameter, with a 7-inch axial hole. There will be eleven thrust-collars $18\frac{1}{2}$ inches in diameter and $1\frac{3}{8}$ inches thick. The diameter of the shaft between collars will be increased to 14 inches. Each shaft is to be carried in bearings forward and abaft the thrust-bearing. No part of the weight is to be taken by the thrust-bearing.

PROPELLER-SHAFTS.

The propeller-shafts will each be in two lengths, $14\frac{1}{2}$ inches diameter. A 9-inch hole will be bored axially through the after section. A 9-inch hole will be bored in the forward section of each shaft from the after end to within 18 inches from the forward end; thence it will be tapered to a diameter of $6\frac{1}{4}$ inches 12 inches from end; it will be further reduced 6 inches from end, and will be threaded and fitted with a $6\frac{1}{4}$-inch steel stud, locked in place. This stud will be reduced to $4\frac{1}{2}$ inches diameter outside the shaft, threaded and fitted with a nut; the stud to extend beyond the nut and to be squared.

The forward section of each shaft will be about 18 feet long, cased with composition $\frac{3}{4}$ inch thick where it passes through the stern-tube up to the outboard coupling. The after section will be 44 feet long, cased with composition $\frac{3}{4}$ inch thick where it passes through the stern-bracket bearing.

The casing will be shrunk and pinned on at the bearings, but between the stern-tube bearings will be carried clear of the shaft.

The different lengths of casing must be made perfectly water-tight at their junctions. The casings must be accurately and smoothly turned to form journals.

The forward end of forward section of shaft will be reduced to a diameter of 14 inches for a length of 12 inches to receive the coupling-sleeve elsewhere specified, and will have three splines for $2\frac{1}{4} \times 2\frac{1}{4}$-inch feather-keys.

The after end of after section will be tapered from $14\frac{1}{2}$ inches to $11\frac{3}{4}$ inches diameter in a length of 2 feet 10 inches to fit the bore of propeller-boss, and will have a spline for a $2\frac{3}{4} \times 2\frac{3}{4}$-inch feather-key.

Abaft this the diameter will be reduced to 10 inches, threaded and fitted with a nut and keeper, the thread being turned off abaft the nut. There will be a water-tight plug in the after end.

There will be at the forward end of the after section of propeller-shaft a cast-steel casing to form a fair water-line from the end of the stern-tube to the shaft. This sleeve will be as tight as possible, finished on the outside, and bored to fit the shaft and couplings. It will be secured to the coupling by six $\frac{3}{4}$-inch naval-brass screws tapped into the coupling-flanges, these screws to have slotted button-heads. The shaft, couplings and casings to be well coated with the same composition as the hull.

SHAFT-COUPLINGS.

All lengths of shafting, except as otherwise specified, will have coupling-flanges forged on.

The different sections of crank-shafts and line-shafts will be coupled to each other by flanges 27 inches in diameter faced 3 inches thick, each with six $3\frac{1}{4}$-inch bolts. The bolt-holes must be equally spaced and drilled accurately to template; the bolt-holes in crank-shafts to be symmetrically placed relatively to the cranks. The coupling-flanges on the eccentric ends of the sections of crank-shafts will have the bolt-holes counterbored $\frac{3}{4}$ inch deep for $4\frac{1}{2}$-inch bolt-heads.

The line and thrust-shafts will be coupled to each other by

flanges 30 inches in diameter, faced $3\frac{3}{4}$ inches thick. These flanges will stand off from each other about $\frac{3}{4}$ inch, and will have for each coupling eight headless bolts, which will be $3\frac{1}{2}$ inches in diameter in the thrust-shaft flange and 3 inches in diameter in the line-shaft flange. The bolts will be tightly set up in the line-shaft flange by wrought-iron nuts. They will fit loosely in the thrust-shaft flange, where they will be lubricated by an approved centrifugal oiling device.

The forward end of the forward section of each propeller-shaft will have a cast-steel sleeve, 2 feet 5 inches in diameter and 13 inches long, slipped on and keyed by three $2\frac{1}{4} \times 2\frac{1}{4}$-inch feathers. The sleeves will be secured in place each by a wrought-iron washer about 15 inches in diameter and $1\frac{7}{8}$ inches thick, fastened to the shaft by the steel stud before specified. This washer will fit a recess in the after end of thrust-shaft. The sleeves will be fluted between the bolt-holes. Each sleeve will be bolted to the after flange of thrust-shaft, which will be 2 feet 5 inches diameter and $3\frac{1}{4}$ inches thick, by six $3\frac{1}{4}$-inch bolts. These bolts will fit the forward ends of the holes in sleeve snugly and will have clearance in the after ends. The two sections of propeller-shaft will have coupling-flanges forged on, $28\frac{1}{4}$ inches diameter and $3\frac{1}{4}$ inches thick, joined by six $3\frac{1}{2}$-inch bolts with nuts permanently locked in place.

The eccentric end of each section of crank-shaft will have a projection 15 inches in diameter and $\frac{1}{2}$ inch long, which will fit a corresponding recess in the adjoining section of crank or line-shaft. The recesses which come together at the junction of the high-pressure and intermediate shafts will be filled with a tightly-fitting steel disc 15 inches diameter and 1 inch thick, with a 7-inch hole in it. There will be similar projections and recesses at the coupling between the two sections of the port line-shaft and at the outboard couplings of the propeller-shafts.

All coupling-bolts will be of forged steel and must fit holes snugly except where otherwise specified. All nuts to be of wrought-iron.

SHAFT PILLOW-BLOCKS.

The line-shaft pillow-blocks will be of cast-iron, faced and bolted to the seatings provided. The caps will be secured each by four $1\frac{1}{8}$-inch wrought-iron bolts. The holding-down bolts will be $1\frac{1}{4}$ inches diameter in $1\frac{3}{8}$-inch holes. Each cap will have a grease-cup cast on, with hinged cover. Each bearing will be lined with anti-friction metal.

The length of line-shaft bearings in forward engine compartment will be 18 inches, that in after engine compartment 24 inches, those in shaft-alleys 16 inches.

There will be a spring-bearing for line-shaft between crosshead slides of air-pump in after engine compartment.

The pedestal of the pillow-block in after engine compartment will have two brackets for supporting the after end of condenser, and two brackets to which the bilge-pump casings will be attached.

THRUST-BEARINGS.

Each thrust-bearing pedestal, of cast-iron, will be firmly bolted to the seating provided, and bored out to receive the lower part of bearing. The bearing will be in two parts, of cast-iron, with white-metal linings. The lower part will be turned to fit the pedestal. The upper part, or cap, will be separated from the bottom by composition distance pieces, and will be fitted in place with six $1\frac{3}{8}$-inch wrought-iron dowel-pins, fitting snugly in holes in the lower part of bearing. The cap will be faced to fit longitudinal recesses in the upper flanges of pedestal, and will be held down by eight $1\frac{1}{4}$-inch wrought-iron bolts, body-bound in pedestal, but with slotted holes in cap. Each cap will have a box cast on top, divided into two parts, one for oil the other for water, the former having a hinged cover. There will be a hole for oil and another for water leading down to each collar and each recess, the white metal being properly channeled. The oil-box will be divided by athwartship partitions.

The end and side walls of the pedestal will extend as high

as the center line of the shaft to form an oil-trough. Inside this trough, both forward and abaft the thrust-collars, will be a composition bearing about 8 inches long for taking the weight of the shaft. These bearings will be adjustable vertically by wedges with regulating-screws.

At each end of each thrust-bearing there will be a divided stuffing-box and gland to prevent the escape of oil. At the bottom of each thrust-bearing there will be a fore-and-aft channel connecting all the bearing recesses; a drain-cock to be fitted at each end.

STERN-TUBE BEARINGS.

Each stern-tube, fitted by the hull contractors, will be finished by the engine contractors as follows: A composition lining, turned to fit the stern-tube, will be inserted from the inner end and secured by a water-tight flange-joint. This lining is to be about $1\frac{1}{4}$ inches thick at forward end and $1\frac{5}{32}$ inches at after end, and will be about 12 feet from face of flange to after end. A prolongation of the forward end of the lining will form the stuffing-box casing. The lining will be bored out at forward and after ends and will be fitted with sections of lignumvitæ. The lengths of the lignumvitæ bearings will be 2 feet 6 inches at each end. The lignumvitæ will be prevented from turning by a composition strip screwed to the tube-lining, and will be properly channeled for the circulation of water. The lignumvitæ must bear on end of grain and will be smoothly and accurately bored out in place to suit the shaft-casing.

STERN-TUBE STUFFING-BOXES.

A composition ring will be inserted in the forward part of each stern-tube lining to form the bottom of the stuffing-box, and will be pinned in place. The gland will be of composition, with naval-brass bolts, fitted with pinion-nuts and spur-ring. The packing space is to be about 6 inches deep and $1\frac{1}{4}$ inches wide. There will be a $1\frac{1}{2}$-inch drain-cock with pipe leading to engine-room bilge.

STERN-BRACKET BEARINGS.

Each stern-bracket is to have a neatly fitting composition lining, about $1\frac{5}{32}$ inches thick, inserted from the after end and secured by a flange-joint. It will be fitted with a lignumvitæ bearing about 4 feet 8 inches long, fitted as in the stern-tubes, and bored out in place to suit the shaft-casing. The lignumvitæ will be held in place at the after end by a flat ring, pinned on, and at the forward end by a conical ring similar to those at after ends of stern-tubes. A light steel sleeve, with interior reinforce ring, will be properly secured to each stern-bracket, and will be so shaped as to make a fair water-line to the propeller-boss.

SCREW-PROPELLERS.

The screw-propellers will be made of manganese bronze. The starboard propeller will be right-handed and the port one left-handed. They will be three-bladed, 14 feet 3 inches in diameter. The bosses will be spherical, about 3 feet 11 inches diameter. Each blade will be secured to the boss by seven $3\frac{1}{8}$-inch naval-brass tap-bolts secured by lock-plates. The recesses for the bolt-heads will be covered by composition plates, held by counter-sunk screws, and finished to form a smooth surface fair with the boss. The pitch to be hereafter determined. The bolt-holes will be slotted to allow the pitch to be varied.

Each boss will be bored to an accurate fit on the tapered end of shaft, and fitted with a wrought-iron $2\frac{3}{4}$ x $2\frac{3}{4}$-inch feather-key.

The developed area of the blades of each propeller will be about 57 square feet. The thickness of the middle of blades at 2 feet 3 inches from center of shaft will be $5\frac{3}{4}$ inches. The thickness at 1 inch from edges of blades will not be over $\frac{5}{8}$ inch; the edges to be made as sharp as possible.

Each propeller will be held on the shaft by a nut, screwed on and locked in place. The shaft-casing is to enter about 1 inch into the propeller-boss and to be fitted water-tight. Each boss will be finished at the after end by a composition

cap, bolted on water-tight. The bosses and caps will be finished all over. The blades will be cast as smooth as possible, and will have any roughness removed. The flanges of the blades are to be turned and faced to fit the recesses in the bosses accurately, and after being secured in place must have the edges made fair.

Three extra blades for each propeller, of such dimensions as may be directed, will be furnished; and will be delivered at such naval station in the United States as may be directed, to be left in store.

CONDENSERS.

The condensers will be cylindrical, 6 feet 1 inch internal diameter, each made in three principal sections;—the middle sections of composition and the others of sheet-brass, Muntz metal or copper.

There will be the following openings in the middle section of each condenser, each with properly faced flanges, viz:

One for main exhaust-pipe, 30 inches diameter;
One for air-pump engine exhaust-pipe, $4\frac{1}{2}$ inches diameter;
Two for air-pump suction-pipes, each $8\frac{1}{2}$ inches diameter;
One for salt-feed pipe, 2 inches diameter;
One for soda-cock;
One for steam-pipe for boiling the water in condenser;
One 15 x 12-inch man-hole at the side of the main-exhaust nozzle;
One 15 x 12-inch man-hole at the bottom. Each middle section will also have two brackets for securing the condenser to the engine-room bulkhead and two supporting brackets.

The tube-sheets will be made of composition or Muntz metal, 1 inch thick, with smoothly-finished holes for the tubes, tapped and fitted with glands for packing the tube ends. The glands will be beaded at outer ends to prevent tubes from crawling, and will be slotted to admit a tool for screwing up. Cotton packing will be used. There will be 4,453 seamless-drawn brass tubes in each condenser, $\frac{5}{8}$ inch outside diameter, No. 20 B. W. G. in thickness. The tubes will be 10 feet

long between tube-sheets and will be spaced $1\frac{5}{16}$ inch between centers. The cooling surface of each condenser will be about 7,276 square feet, measured on the outside of the tubes.

The end sections of each condenser will be single-riveted to the middle section, and will have flanges riveted on to connect with the tube-sheets. The longer section of each condenser will be strengthened at its middle by a composition ring riveted on. The longitudinal seams will be double-riveted. The tube-sheets will be secured to the flanges of the shell by naval-brass collar-bolts, which will also be used for fastening the circulating-water chests.

The chest for entrance and exit of circulating water will be made of composition, with a division-plate in the middle and with a man-hole in each compartment. The inlet and outlet-nozzles will each be 17 inches in diameter of opening.

The water-chest at the other end of the condenser will be built up in dished form, as shown, with composition flange, man-hole frame, and man-hole plate, and with a sheet-brass body. There will be two stay-bolts to connect the man-hole frame to the tube-sheet.

There will be six braces of rolled Muntz metal connecting the tube-sheets, each $\frac{5}{8}$ inch in diameter, and each passing through a stay-tube 1 inch external diameter and No. 10 B. W. G. in thickness.

Baffle-plates of brass, Muntz metal or copper will be fitted, as shown, to direct the steam over all the tubes. Plates will be provided for supporting the tubes and to act also as baffle-plates.

On top of the horizontal baffle-plate there will be a vertical dividing-plate, 3 inches high, extending from end to end. At each end of this horizontal plate there will be a vertical plate, 3 inches high, extending 24 inches each side of the middle, with holes for the tubes to pass through. At the bottom of each condenser there will be a vertical dividing-plate extending from the shell to the lower row of tubes, to insure an equal supply of water of condensation to each air-pump suction-pipe.

In front of the main exhaust-nozzle, above the tubes, will be a deflecting-plate, supported as shown. This plate to be perforated only where not in the direct line of the incoming steam.

Each condenser will be supported at the end nearest the air-pump by two prolongations of the tube-sheet, which will be secured by angles bolted to brackets cast on the top of the air-pump casing. Each condenser will be supported near the middle by two 3-inch wrought-iron stanchions bolted to the facings provided for them, as shown, on the middle section of the condenser, and resting upon and bolted to seatings provided for the purpose. The after condenser will be supported at its after end by two prolongations of the tube-sheet, which will be secured by angles to brackets on the line-shaft pillow-block underneath. The forward condenser will be supported at its forward end by a $\frac{1}{2}$-inch iron plate, properly stiffened by angles, which will be built up from the inner bottom of the vessel and bolted to a prolongation of the tube-sheet.

Each condenser will be connected by three wrought-iron plate-brackets to the fore-and-aft engine-room bulkhead,—the end brackets to be bolted to the tube-sheets and the middle bracket to a facing provided, as shown, on the middle division of the condenser.

A 2-inch salt-feed pipe, with a spray in the exhaust-passage, will be fitted to each condenser. A copper tank, pipe and cock to be provided for admitting an alkaline solution into the condenser; this pipe to connect with the salt-feed spray; the tank to be of at least 10 gallons capacity and conveniently placed. A $1\frac{1}{2}$-inch branch from the auxiliary steam-pipe will lead to the bottom of the condenser for cleaning the tubes by boiling.

The salt-feed pipe of each condenser will be connected with the delivery side of the engine-room auxiliary-pump. Drain-cocks will be provided, with pipes leading to the bilge.

All parts of the condensers, except as otherwise specified, will be made of composition. All bolts to be made of naval brass. All bolts for securing flanges of pipes and man-hole plates will be standing bolts, and will, wherever possible, be

screwed into the condenser-plates with heads inside. The condensers must be perfectly tight all over and be so proved after being secured in place.

AIR-PUMPS.

There will be a double double-acting horizontal air-pump, driven by a two-cylinder vertical inverted cylinder compound engine, for each propelling-engine. The engine will drive a crank-shaft, carried in four bearings, with cranks at right-angles and with a fly-wheel at each end. The air-pump connecting-rods will be connected to the same crank-pins as the engine connecting-rods—the former having forked ends and the latter single ends.

All parts of the pumps, except as otherwise specified, will be made of composition. Each air-pump will have a piston working in a cylinder of 20 inches bore. The stroke will be 18 inches. The pump-pistons will be cast hollow, and must be tested for perfect tightness. Each piston will have a wearing surface of lignumvitæ, as shown, bearing on end of grain. The piston packing to be made of hemp, set up by a follower with locked nuts. The pump piston-rods and follower-bolts will be made of rolled phosphor-bronze or aluminum-bronze. There will be fifteen foot-valves and fifteen delivery-valves at each end of each pump, all 4 inches in diameter, made of the best hard rubber or of such other material as may be approved. Each valve will be held in place by a guard and a spiral spring of phosphor-bronze. The valves and guards must be easily removable, and held firmly in place. The valve-seats will be made separately from the pump-casings and will be bolted in place. The foot-valve seats will be placed in inclined positions at the sides of the pumps, and the delivery-valves horizontally at the highest parts of the pump-chambers. There must be no pockets in the pump-chambers, underneath the delivery-valves, where vapor can lodge. The gratings of the valve-seats must be so arranged that the clear opening of each valve shall be at least $6\frac{1}{4}$ square inches. The pump-barrels will be provided with working linings. The pump-

casings and bonnets will be well ribbed. The bonnets will be provided with jack-bolts and eye-bolts. Each air-pump will have two suction-nozzles, each $8\frac{1}{2}$ inches in diameter. Each of these nozzles will be connected by a copper pipe and straightway-valve to the corresponding nozzle on the bottom of the condenser.

Each hot-well will be formed to act as an air-chamber. Each air-pump will have two outlets, each 6 inches in diameter, connected by a copper pipe, a prolongation of which will lead to the feed-tank.

The engine-cylinders will be 12 and 19 inches bore respectively. Each cylinder will be supported at the back by a cast-iron frame forming the cross-head guide and at the front by a wrought-iron column. The cylinders and bed-plates will be bolted together as shown. The high-pressure and low-pressure valve-chests of each engine will be connected by a copper pipe $3\frac{1}{2}$ inches diameter. The pistons and cylinder-covers will be made of cast-steel. The crank-shafts, piston-rods and connecting rods will be made of forged steel. The pistons will be fitted with cast-iron packing-rings. The cross-heads will have cast-steel slippers lined with white metal, working in cast-iron guides with composition guards. The crank-shafts will be 4 inches diameter, with bearings 5 inches long. Each crank-pin will be $4\frac{1}{2}$ inches diameter and 7 inches long. Each cylinder will have a slide-valve, worked through gear, as shown by a pin carried on a plate so secured to the fly-wheel that its position may be varied.

Each engine will take steam from a branch of the main steam-pipe, with a stop-valve having a hand-wheel at the working-platform, and will exhaust by a special pipe into the condenser.

The air-pump engines will be connected to the pumps by cast-iron frames, which will form the air-pump cross-head guides. The engines, pumps and frames will be securely bolted to the seatings provided for them. The air-pump piston-rods will be continued through the cylinder-covers, and will be connected by keys to the main bilge-pump plungers.

Each air-pump, together with its condenser, must maintain a vacuum of within 4 inches of mercury of the weather barometer with the propelling-engines at full power under forced draught.

MAIN BILGE-PUMPS.

The main bilge-pumps will be of the plunger type, two in each engine compartment. These pumps will be placed behind the air-pumps, and will be worked by prolongations of the air-pump piston-rods, to which their plungers will be keyed. Each plunger will be 6 inches in diameter, cast hollow, as light as possible, and will be tested for tightness. The pumps will be made in all parts, except as otherwise specified, of composition. There will be four foot-valves and four delivery-valves for each pump. The valves will be 3 inches in diameter, made of phosphor-bronze $\frac{1}{16}$ inch thick, and will each be secured by a guard with a phosphor-bronze spiral spring. The valves must be so arranged as to be easily overhauled when the pumps are running. Each pump will have a copper air-chamber on the delivery side. The suction-pipe of each pump will be $4\frac{1}{2}$ inches bore and the delivery-pipe $3\frac{1}{2}$ inches bore. There will be a non-return valve in each delivery-pipe, so arranged with a screw-stem that it may be lifted off its seat, but may not be screwed down. The bilge-pumps in the after engine compartment will be secured to the line-shaft pillow-block under the after end of the condenser. The bilge-pumps in the forward engine compartment will be supported by a continuation of the air-pump seatings.

CIRCULATING-PUMPS.

There will be a centrifugal circulating-pump for each condenser, each driven by an inclosed three-cylinder engine. Each pump must be capable of discharging 8,000 gallons of water per minute from the bilge. The pumps will be made of composition, except as otherwise specified. Each pump-casing will be made in two parts, divided, the upper part with conveniences for handling. The suction-nozzle will form a

support for the pump and will be securely bolted to its seating. This nozzle will have a 17-inch opening for sea suction and a 15-inch opening for bilge suction. The pump-runners will be 42 inches in diameter, smoothly cored, finished on the outside, and perfectly balanced. The shafts will be of phosphor-bronze. The bearings will consist of sections of lignum-vitæ on end of grain, dovetailed into composition split-sleeves, which will be well secured against turning. The stuffing-box glands will be each in two parts. There will be an air-cock at the top of the pump-casing and a drain-cock at the bottom. The pump-casings must be made as light as possible consistent with strength, and must be smoothly cored, with easy bends wherever the direction of the flow of water is changed.

CIRCULATING-PUMP ENGINES.

The circulating-pump engines will be of the inclosed three-cylinder single-acting type, of approved pattern,—each of sufficient power to secure the results above specified. The engine-valves must be of either the slide or piston type. All lubrications of the engines to be automatic.

CIRCULATING-PUMP CONNECTIONS.

Each circulating-pump will be fitted with pipes and valves to draw from the sea or engine-room bilge, and will deliver into the condenser; also direct to the outboard-delivery pipe by a 17-inch pipe connecting inlet and outlet of condenser. This pipe and the inlet-pipe to condenser each to have a damper-valve of approved pattern.

The injection and delivery-pipes will be 17 inches internal diameter.

There will be damper-valves in the pipes leading from the sea and from the bilge in each engine compartment. These valves will be so connected by levers that when one is shut the other is open; and both will be worked by a long lever well above the floor-plates.

MAIN INJECTION-VALVE.

There will be a 17-inch straight-way screw main injection-valve in each engine compartment. It will connect with the sea by a conical steel tube through the double-bottom in after engine compartment, between frames 71 and 72, and between second and third longitudinals from keel. In the forward engine compartment it will connect with the sea just above the double-bottom, between frames Nos. 57 and 58. There will be a strainer in each pipe at the ship's side.

The valve in forward compartment will be contained in a trunk under the coal-bunker, by which it may be reached from the engine-room. It will be worked from the engine-room by bevel gearing.

There will be a $1\frac{1}{2}$-inch steam-pipe leading from the auxiliary steam-pipe to the injection-pipe outside of injection-valve at each end of pipe.

BILGE-INJECTION VALVE.

There will be a 17-inch bilge-injection pipe connected with each circulating-pump; each pipe to have a 17-inch non-return valve, that can be lifted from its seat by an outside screw-stem.

MAIN OUTBOARD-DELIVERY VALVES.

There will be in each engine compartment a main outboard-delivery valve of the same size and type as main injection-valves.

The valve in after engine compartment will connect with a steel pipe about $\frac{3}{8}$ inch thick, passing through the double-bottom between frames 74 and 75 and between the second and third longitudinals.

The valve in forward engine compartment will connect with the ship's side, between frames 57 and 58, just below the protective-deck; the pipe and valve to be inclosed in a trunk to be worked from the engine-room by bevel gearing. Zinc protecting-rings to be fitted to valves.

FEED-TANK.

There will be a feed-tank in forward engine-room of about 120 cubic feet capacity. It will be made of $\frac{5}{16}$-inch wrought-iron—utilizing the engine-room bulkhead for one side of the tank. It will be stayed internally as may be directed. It will have at least 140 cubic inches of rolled zinc plates about $\frac{1}{2}$ inch thick, suspended from the braces. The straps suspending the zinc plates, and the braces where the straps come in contact, are to be filed bright before fitting; the joints to be then well painted on the outside, or be made water-tight in other approved manner. On top of the tank there will be a filter, into which the water from the air-pumps will be delivered. The filter will be provided with sponges and will be so arranged that the sponges are readily accessible. The tank will have a man-hole with bolted cover. It will have a glass water-gauge with suitable guards, shut-off cocks and drain-cocks. The tank and filter will have the following pipe connections: A discharge-pipe from each air-pump; an overflow-pipe leading to bilge, but so arranged that any water passing down it may be seen—this pipe to have a non-return valve; a suction-pipe to forward feed-pump, with valve; a suction-pipe to after feed-pump, with valve; drain-pipes from traps, as elsewhere specified; a vapor-pipe, 6 inches diameter, of copper, No. 16 B. W. G. The vapor-pipe will lead up the engine-room hatch and discharge above the level of the awnings, where it will have a suitable hood; or it may be led into the main escape-pipes. Each feed-pump suction will be provided with a valve operated by a copper float in the feed-tank, so arranged that it will allow no air to enter the feed-pipes. There will be a small supplementary feed-tank in the after engine-room above the level of the air-pump, into which the air-pump will discharge. From this tank a 6-inch vapor-pipe will lead as specified for the forward compartment, and a 4-inch water-pipe will lead to feed-tank in forward engine-room. The main feed-tank will form a support for the forward end of forward condenser.

BOILERS.

There will be four cylindrical, double-ended, horizontal return-tube boilers 14 feet 8 inches greatest diameter and 19 feet 2 inches mean length over heads. Each boiler will have three corrugated furnaces 42 inches in least internal diameter. The boilers will be divided into two groups, each in a separate water-tight compartment.

BOILER MATERIAL.

All material used in the construction of the boilers, except the tubes, rivets and circulating-plates, is to be open-hearth steel. The tubes are to be of wrought-iron or steel of the best quality. The rivets to be of either open-hearth or Clapp-Griffiths' steel. All material is to be tested as elsewhere specified.

BOILER-SHELLS.

The shell of each boiler will be $1\frac{1}{32}$ inches thick, made in three rings, each ring of three plates.

BOILER-HEADS.

Each head of each boiler is to be made of three plates; the upper plate $1\frac{1}{32}$ inches thick, the middle plate forming the front tube-sheet will be $\frac{3}{4}$ inch thick, and the lower plate $1\frac{1}{16}$ inch thick. The lower plates of front heads to be flanged inwardly at each furnace and all plates flanged inwardly at circumference. The heads will be curved to a radius of 4 feet $2\frac{3}{4}$ inches at the top, as shown in drawing.

BOILER TUBE-SHEETS.

The front and back tube-sheets will each be $\frac{3}{4}$ inch thick. The tube-sheet will be drilled for stay-tubes and plain tubes, as shown, and tapped in place for the stay-tubes. All tube-holes to be slightly rounded at edges. Each pair of tube-sheets must be accurately parallel.

BOILER-TUBES.

All boiler-tubes are to be of the best quality of wrought-iron or steel, lap-welded, 6 feet 9 inches long between tube-sheets. There will be about 114 stay-tubes and 416 ordinary tubes in each end of each boiler. All tubes will be $2\frac{1}{4}$ inches external diameter. The ordinary tubes will be No. 12 B. W. G., and will be swelled to $2\frac{5}{16}$ inches external diameter at the front ends. They will be expanded in tube-sheets at both ends and beaded over at combustion-chamber ends. The stay-tubes will be No. 6 B. W. G., upset at both ends to $2\frac{3}{8}$ inches external diameter, leaving the internal diameter as in the remainder of the tube, then swelled at the front end to $2\frac{1}{2}$ inches external diameter. They are to be threaded parallel at combustion-chamber end and tapering at front end to fit threads in tube-sheets. They are to be screwed into the tube-sheets to a tight joint at front end. The combustion-chamber ends will be made tight by expanding and beading. All expanding is to be done by approved tools. The tubes will be spaced $3\frac{1}{4}$ inches from center to center vertically and $3\frac{1}{2}$ inches from center to center horizontally.

COMBUSTION-CHAMBERS.

There will be a separate combustion-chamber for each furnace. They will be made of $\frac{1}{2}$-inch plates except the tube-sheets, which will be as before specified. The backs of the combustion-chambers will be rounded at the top to a radius of 2 feet 3 inches, as shown in the drawing. The plates will be flanged where necessary, and all parts joined by single-riveting. The holes for screw stay-bolts in plates of combustion-chambers and shells are to be drilled and tapped together in place.

BOILER BRACING.

The bracing of each boiler will be as follows:
Four 2-inch and five $2\frac{1}{2}$-inch through-braces in one horizontal row, spaced 14 inches center to center, connected to both heads just below the beginning of the curve at the upper part.

Two $1\frac{9}{16}$-inch through-braces connecting the heads above the lower man-holes. All through-braces will have raised threads on ends, with inside and outside nuts at both heads.

There will be at each end of each boiler four $1\frac{9}{16}$-inch braces with raised thread and inside and outside nuts at head ends, and with the other ends flattened and riveted to boiler-shell as shown. These braces to be at bottom of lower man-holes.

All diagonal braces will have their nuts set up on beveled wrought-iron washers riveted to both sides of heads of boilers.

The backs of combustion-chambers will be stayed by $1\frac{3}{16}$-inch screw-stays, spaced 8 inches vertically and horizontally, screwed into the sheets of the opposite chambers and nutted. The sides of combustion-chambers will be stayed in the same manner to the boiler-shell.

The top of each combustion-chamber will be stiffened by $3\frac{1}{2} \times 3\frac{1}{2} \times \frac{5}{8}$-inch angles, as shown in the drawing, riveted to the curved plate as shown. Each pair of opposite angles will be connected by a plate-stay.

The bottoms of combustion-chambers will be stiffened by angles as shown in drawings.

RIVETED JOINTS.

The longitudinal joints of boiler-shells will be butted, with $\frac{5}{8}$-inch straps on the inside and $\frac{3}{4}$-inch straps on the outside, and treble-riveted. All circumferential joints to be lapped and double-riveted. All joints in furnaces and combustion-chambers will be single-riveted and lapped or butted as shown in the drawing. All rivets to be of steel. Edges of all plates to be planed. The proportions of all joints to be as approved. All rivet-holes to be drilled in place after bending. Hydraulic riveting to be used where possible. In parts where hydraulic riveting cannot be used the rivet-holes are to be coned and conical rivets used. All seams to be calked on both sides in an approved manner. Longitudinal seams must break joints as shown.

BOILER MAN-HOLE AND HAND-HOLE PLATES.

There will be three man-holes in each boiler-head. Each man-hole will have a raised steel frame riveted on the head; the upper man-hole to be 12 x 15 inches and the lower ones at least $10\frac{1}{2}$ x 15 inches. The man-hole plates will be dished—made of cast-steel. Each plate will be secured by two wrought-iron dogs and two $1\frac{1}{4}$-inch studs with square nuts. Each man-hole plate will have a convenient handle. All plates, dogs and nuts to be indelibly marked to show to what holes they belong. There will be in each head of each boiler four 6 x 8 hand-holes, with plates fitted the same as for man-holes, except that each will have but one bolt.

STOP-VALVE NOZZLE.

There will be a cast-steel or bronze nozzle with a flange corresponding to the stop-valve flange, and another flange to fit the curvature of the upper part of boiler. The nozzle to extend into the boiler and have the dry-pipe bolted to it.

FURNACES.

Each furnace is to be in one piece, $\frac{1}{2}$ inch thick, and corrugated; 3 feet 6 inches least internal diameter, and 3 feet 10 inches greatest external diameter. They must be perfectly circular in cross-section at all points. They will be riveted to flanges of front heads and will be flanged and riveted to combustion-chamber plates. The corrugations of adjacent furnaces are to alternate.

FURNACE-FRONTS.

The furnace-fronts will be made of cast-iron frames, with wrought-iron outer plates bolted on. The dead-plates of each front will be made of cast-iron, neatly fitted to the frame, and secured by two $\frac{1}{2}$-inch bolts with countersunk heads. The dead-plates must be interchangeable, and when in place must come flush with the furnace-fronts.

There will be at each side of the furnace-door an opening in the bottom of the door-frame about $3\frac{1}{2} \times 4\frac{1}{2}$ inches, to admit air from the ash-pit into the hollow of the frame.

Ports will be cored in the inner wall of the door-frame to admit air from the hollow of the frame into the space between inner and outer plates of the furnace-door. Each furnace-front will rest on two lugs secured to the furnace-flue, and will be further secured by bolts screwed into the boiler-heads; it will have allowance for expansion around its periphery, and will have a flange standing off from the end of the boiler-head about one-quarter inch, the intermediate space being packed with suitable non-combustible material. A projection will be cast on each frame to hold the upper hinge of furnace door. The door-opening will be about 30 inches wide and 17 inches high at middle. There will be a beading on the inside of the door-frame in wake of the furnace-door lining to make the clearance around the door lining.

FURNACE-DOORS.

The furnace-doors will be made with inner and outer plates of $\frac{1}{4}$-inch wrought-iron. The outer plate will be flanged one inch inwardly around its periphery, the edge of the flange to be made a neat fit to door-frame. The inner plate will stand off from the outer plate about $4\frac{1}{4}$ inches, and will be secured to it by eight $\frac{3}{8}$-inch socket-bolts with nuts outside. It will be made to have as little clearance as practicable when closed. It will be perforated as may be directed. There will be three hinges for each door; the two lower hinges will have the working parts of wrought-iron, each bolted to the outer edge of door by two bolts; the upper hinge will have a $\frac{3}{4}$-inch rod passing through an eye riveted to outer shell of door with a nut for taking up the sag of the door. The standing part of the hinges will be of wrought-iron, each with a double eye; the two lower ones will be riveted to the outer plate of furnace-front, and the upper one bolted to the projection made for the purpose on the furnace-front frame. The pintles of all hinges

will consist of $\frac{5}{8}$-inch rivets dropped into place. The latches, with their straps, will be of wrought-iron, bolted to the doors; the catches will be of wrought-iron, riveted to furnace-fronts.

ASH-PIT DOORS.

The ash-pit doors will be made of $\frac{1}{8}$-inch wrought-iron with a continuous $1\frac{3}{4}$ x $\frac{3}{8}$-inch channel-iron worked around the inner edge. This channel will have its edges lipped over and will be filled with asbestos or other approved non-combustible packing. Each door will be fastened in place by four wrought-iron buttons bolted to lugs on the walls of the air-duct, each button setting up on a wedge riveted to the door. The top of each door will be stiffened by an angle-iron riveted on the inside. Each door will have two wrought-iron handles, two wrought-iron beckets to fit hooks on uptake-doors, and an eye for slinging by. The upper parts of the handles will be formed into hooks for hanging the movable portions of the air-ducts.

BRIDGE-WALLS.

The bridge-wall frames are to be made of cast-iron, well ribbed, fastened to furnaces so as to be readily removed. The bridge-walls will be carried back to meet the backs of combustion chambers, so that there shall be no part of the cast-iron above the grate-bars unprotected by fire-brick, and so that no ashes can get into the spaces back of bridge-walls. They will be finished with fire-brick or such other refractory material as may be approved.

GRATE-BARS AND BEARERS.

The grate-bars in wake of the furnace-doors will be $1\frac{1}{4}$ inches square, of wrought-iron, of revolving pattern. They will be swaged down to circular cross-section where they pass through the furnace-fronts. The grate-bars at the sides of the furnaces will be of cast-iron, in four lengths; those nearest the sides of the furnace-flues will be shaped to fit the corrugations and bolted to straight bars with thimbles between to form

air-spaces. Straight bars to be made single. Each grate will have a total length of 7 feet.

The bearers will be of wrought-iron. The bearer nearest the furnace-front will be fitted to support the fixed grate-bars only. The three intermediate bearers in each furnace will be double—one-half of each fitted for both revolving and fixed bars, the other half for fixed bars only; the bearer nearest to or forming part of the bridge-wall will be fitted to support all bars. All bearers will be supported by wrought-iron lugs bolted to furnace-flues.

LAZY-BARS.

A lazy-bar with the necessary lugs will be fitted in the front of each ash-pit.

ASH-PANS.

Ash-pans of $\frac{1}{4}$-inch wrought-iron, reaching from the fron of furnace-flue to bridge-wall, are to be fitted to all furnaces

CIRCULATING-PLATES.

Each boiler will have circulating-plates fitted at each side of each nest of tubes. They will be of wrought-iron $\frac{1}{16}$ inch thick, in sections as shown in drawing. Each section will have two clips at upper and two at lower end for supporting it from the stay-tubes, as shown. The plates will be well painted all over with two coats of brown zinc and oil.

AIR-DUCTS.

From each of the two blowers in each fire-room a wrought-iron trunk will lead horizontally, the two trunks joining in a vertical trunk leading to the space underneath the fire-room floor-plates. Each trunk to be fitted with a damper which can be quickly closed in case its blower is stopped. The fire-room floor-plates will be bolted to the ledges of the duct under fire-room floor, the joints to be air-tight. The sides of

the air-duct under fire-room floor will be formed of $\frac{3}{16}$-inch wrought-iron plates, worked from front heads of boilers to inner skin of ship, and bolted to angles on each in such a manner as to be easily removed for repairs to boilers. From this main air-duct a duct of $\frac{3}{16}$-inch plates will lead to each ash-pit, and will be fitted with lugs and buttons for securing the ash-pit doors. The fronts of the air-ducts opposite the lower man-holes to be easily removed. These removable sections will be fitted with handles, eyes, and beckets similar to those on ash-pit doors.

In each of these ducts a pivoted damper will be so fitted as to be easily removable for repairs, and provided with a lever outside the air-duct for opening and closing. The damper-gear to be so fitted that the amount of opening of the dampers can be easily regulated.

The dampers must be made to work easily, and when closed must be practically air-tight.

UPTAKES.

The uptakes will be made of wrought-iron, No. 8 B. W. G., built on angles and bolted to boiler-heads and shells. They will be bolted to the lower plates of smoke-pipe with slotted holes to allow for expansion.

The uptakes are to be carried up clear of the ends of the boiler-stays. There will be a fore-and-aft division-plate in each uptake to keep the gases from the boilers separate.

The space between the plates of the uptake will be filled with an approved incombustible non-conducting material.

UPTAKE-DOORS.

The uptake-doors will be made of double shells of iron of the same thickness as uptakes. The space between the shells will be filled with the same non-conducting material as in uptakes.

The hinges and latches will be made of cast-steel. Each

door will have two hooks for hanging the ash-pit doors and a hook for a rope for hoisting the same.

Each door will also have an eye near its top for handling.

SMOKE-PIPES.

There will be two smoke-pipes: one for each pair of boilers 7 feet 7 inches by 5 feet feet 7 inches internal diameter, and 55 feet high above the grate-bars of lower furnaces.

The fore-and-aft division-plate of each uptake will be continued in the smoke-pipe up to about the level of the main deck. There will also be an athwartship division-plate.

The pipes will be made of wrought-iron, No. 7 B. W. G. for the lower half, and No. 9 for the upper half. The lower part of each pipe will be stiffened by angle-irons. Above the beginning of the oval part there will be internal cross-stays of $1\frac{1}{8}$-inch iron with T-heads riveted to angles. Each pipe will be finished at the top by an angle-iron to which the stay-shackles will be secured, and by a hood covering the casing, to which will be secured shackles for slinging painters. Each pipe will be strongly stayed by guys and turn-buckles of approved dimensions and pattern. All joints will be butted and strapped. The pipes will be supported in an approved manner,—all the work being done by the contractor for the machinery.

From its junction with the uptakes to about 6 inches below the hood at top each smoke-pipe will be surrounded by a casing, leaving an annular space of 5 inches. The casing will be made of iron, No. 12 B. W. G., and strengthened by angle-irons. It will be butted and strapped, flush-riveted on the outside, and open top and bottom. It will be stayed to the pipe, and will be finished with a half-round iron at top. There will be doors through this casing and through smoke-pipe about on a level with the main deck.

Above the smoke-pipe hatch an oval iron casing, No. 12 B. W. G., about 11 feet 7 inches by 9 feet 7 inches, will ex-

tend and will be finished by half-round iron. About 1 foot above this there will be a hood carried by the casing.

There will be a ladder on the outside of each pipe, reaching to the top of the forward part.

SMOKE-PIPE COVERS.

Each smoke-pipe will have a permanently-fixed cover made of wrought-iron, No. 11 B. W. G., built on angles in a slightly-dished form and supported by angles riveted to the smoke-pipe. The cover will be placed about 30 inches above the top of the smoke-pipe so that it will not interfere with the exit of the gases, and will overlap the smoke-pipe about 18 inches all around.

BOILER-SADDLES.

Each main boiler will rest in three saddles, which will be built in and form part of the hull. It will be secured to the saddle by bolts of approved dimensions, screwed into the boiler-shell with heads inside. The bolts are to pass through oval holes in the saddles to allow for expansion, and will have washers and nuts on the under side.

MAIN BOILER ATTACHMENTS.

Each main boiler will have the following attachments, viz:
One steam stop-valve;
One dry-pipe;
One main feed check-valve with internal pipe;
One auxiliary feed check-valve with internal pipe;
One surface blow-valve with internal pipe;
One bottom blow-valve with internal pipe;
Two safety-valves;
Two steam-gauges;
Three glass water-gauges;
Eight gauge-cocks;
Two sentinel-valves;

One salinometer-pot;
Two drain-cocks;
One air-cock;
Two hydrokineters or equivalent;
One cock with thread for the attachment of a syringe.

All external fittings will be of composition unless otherwise directed. No fittings are to be screwed into the boiler-plates, but will be flanged and through-bolted or attached in other approved manner. All cocks, valves and pipes will have spigots or nipples passing through the boiler-plates. All internal pipes are to be of brass, No. 14 B. W. G., and must touch the plates nowhere except where they connect with their external fittings. The internal feed and blow-pipes will be expanded in the holes in boiler-shells to fit the nipples on their valves, and they will be supported where necessary in an approved manner. The stems of all valves on boilers are to have outside screw-threads. The internal feed and blow-pipes are to be arranged to come between the corrugations of furnaces.

BOILER STOP-VALVES.

There will be a $10\frac{3}{4}$-inch self-closing main stop-valve, with horizontal spindle, on the forward end of each boiler. The valve to be bolted to the projection of the steam-pipe from the boiler.

There will be a nozzle, with suitable flanges, at right angles to the axis of the valve, to which the steam-pipe will be bolted.

There will be a bracket cast on each valve-chest, to be bolted to boiler-head for additional stiffness. A screw-sleeve, with suitable hand-wheel, will be fitted for closing the valve.

DRY-PIPES.

Each boiler will have a brass dry-pipe, No. 14 B. W. G., tapering from $10\frac{3}{4}$ inches to 8 inches internal diameter, with closed end. It will be placed as high as practicable and suitably supported. It will have 374 slits at the top, 4 inches long, $\frac{1}{8}$ inch wide, $\frac{1}{4}$ inch apart, in groups, as shown. The

dry-pipes and their nozzles to be so made that there can be no possibility of water collecting in any part, but that any water entering the dry-pipes will be drained into the steam-pipes to be carried to the separators.

FEED CHECK-VALVES.

The main and auxiliary check-valves will each be $2\frac{3}{4}$ inches in diameter. They will be fixed on the forward ends of the forward boilers and the after ends of the after boilers, and fitted with internal pipes leading above the tubes and pointing downward in the water-spaces between the nests of tubes.

BOTTOM BLOW-VALVES.

The bottom blow-valves will be $2\frac{1}{4}$ inches diameter, placed at ends of boilers opposite the check-valve ends, but with approved gear for working from the fire-rooms, where the feed-pumps are situated. The internal pipes will lead to bottoms of boilers.

SURFACE BLOW-VALVES.

The surface blow-valves will be $2\frac{1}{4}$ inches diameter and placed at same ends of boilers as the bottom blow-valves, but worked from the opposite ends in an improved manner. Each internal pipe will lead to a dish about one inch above the highest heating surface.

SAFETY-VALVES.

Each boiler is to have two $5\frac{1}{4}$-inch spring-pop safety-valves in one case, bolted to the after end of each boiler, and to be clear of uptake. Each set of valves is to be fitted with approved gear for lifting from the main deck as well as from fire-rooms, the gear to work independently of each other. The toes for lifting the valves will be stepped so that the valves will be lifted in succession. All joints in the lifting-gear to be composition-bushed. Each valve is to be set at 138 pounds per square inch, and is to be furnished with washers for the reduction of pressure to 93 pounds by decrements of 5 pounds.

Each washer is to be marked to show its place. A drain-pipe will be attached to each safety-valve case at the lowest part, and will lead to the bilge.

SENTINEL-VALVES.

Each boiler will have, upon each end, a sentinel-valve of $\frac{1}{2}$ square inch in area. It will have a sliding weight on a notched lever graduated to 150 pounds pressure.

MAIN BOILER STEAM-GAUGES.

There will be two Lane's improved spring steam-gauges on each main boiler with $8\frac{1}{2}$-inch dials. The gauge on the feeding end of each boiler will be graduated to 230 pounds and that on the other end to 160 pounds. Each gauge is to have an independent connection to its boiler, and will be fitted with a three-way cock and a coupling for attachment of a test-gauge.

MAIN BOILER WATER-GAUGES.

Each main boiler will have three glass water-gauges of approved pattern, two on the feeding end and one at the other end. Each gauge will be placed at the side of the boiler, and will have $1\frac{1}{2}$-inch pipes leading to top and to near bottom of boiler, with a valve in each close to boiler. The shut-off and blow-out cocks are each to have at least $\frac{1}{2}$ inch clear opening, and are to be packed cocks, with levers and rods for working from fire-room. The glasses are to be about 16 inches in exposed length, with the lowest exposed part about one inch below the highest heating surface. The glasses to be well protected. A brass index-plate, with letters and arrows cast in relief, is to be fixed close to each gauge-glass to show the height of the top of combustion-chamber. The blow-out cocks will have drain-pipes leading to bilge, with union joints.

GAUGE-COCKS.

There will be four asbestos-packed gauge-cocks of approved

pattern on each end of each boiler, with rods and levers for working from fire-room. Each cock is to have an independent attachment to the boiler. They will be spaced about 6 inches vertically, the lowest one being about 4 inches below the highest heating surface. Each set of cocks to have a drip-pan and a drain-pipe leading to bilge.

SALINOMETER-POTS.

There will be a salinometer-pot of approved pattern connected to each boiler, and placed in the feeding fire-room where directed.

BOILER DRAIN-COCKS.

Each main boiler will have at each end a 1-inch drain-cock of approved pattern, or a screw-plug.

BOILER AIR-COCKS.

Each main boiler will have a $\frac{1}{2}$-inch air-cock at its highest part, with a $\frac{5}{8}$-inch copper pipe leading to bilge.

HYDROKINETERS.

There will be fitted to each main boiler two Weir's hydrokineters, or other approved devices, for circulating the water in the boiler while raising steam. Each of these will be fitted where directed, and will have a stop-valve close to boiler. They will take steam from the auxiliary steam-pipe, with stop-valve in fire-room.

ZINC BOILER PROTECTORS.

Each main boiler will have thirty-six rolled zinc plates 12 x 6 x $\frac{1}{2}$ inch. Each plate will be bolted to a wrought-iron strap, which will be clamped to the stays. Each strap will be filed bright where in contact with zinc and stay, each stay being also filed bright at contact point. After being bolted in place the outside of the joints will be made water-tight by paint or approved cement.

BOILER CLOTHING.

After the boilers are tested and painted they will be covered on tops and sides, and on ends where required, by $\frac{1}{16}$-inch galvanized wrought-iron plates, lapped and bolted to angles, which will be held in place by $\frac{1}{2}$-inch bolts tapped part way into the boiler-plates. There will be a space of not less than 1 inch between boilers and covers, which will be filled with approved incombustible non-conducting material securely held in place.

AUXILIARY BOILER.

There will be one cylindrical horizontal return fire-tube boiler for auxiliary purposes, placed where shown on drawings. It will be made of the same material and similar parts will be finished and fitted in the same manner as in the main boilers, except as specified below.

The boiler will be 8 feet external diameter and 8 feet long. The shell will be in one length of $\frac{11}{16}$-inch plates, with butted and double-strapped longitudinal seams. The heads, except the lower plate of the front head, will be $1\frac{1}{4}$ inch thick. The tube-sheets, front and back, will be $1\frac{3}{16}$ inch thick. The back and sides of combustion-chamber will be $\frac{1}{2}$ inch thick. There will be one corrugated furnace, 3 feet 3 inches least internal diameter, of $\frac{1}{2}$-inch plate. There will be thirty-two stay-tubes and eighty-two ordinary tubes, $2\frac{1}{4}$ inches external diameter, secured in the same manner and of the same thickness as in main boilers, spaced $3\frac{1}{2}$ inches between centers horizontally and $3\frac{3}{4}$ inches vertically. The tubes will be 5 feet 8 inches long between tube-sheets. The back-sheet of the combustion-chamber will be rounded off to form the top. The back-sheet of the combustion-chamber will be stayed to the back-head of the boiler by $1\frac{1}{4}$-inch stays spaced 7 inches vertically and horizontally, screwed into both plates and nutted; also by one horizontal row of heel-braces, spaced as above, connecting the curved top of combustion-chamber to the back-head. The sides and bottom of combustion-chamber will be stayed to the boiler-shell by two rows of $1\frac{1}{4}$-inch screw-stays, fitted

Four gauge-cocks;
Two steam-gauges;
One sentinel-valve;
One drain-cock;
One salinometer-pot;
One air-cock.

All of the above to be of the same pattern as for main boilers.

AUXILIARY FEED-TANK.

A feed-water tank of about 100 gallons capacity will be fitted in the auxiliary boiler fire-room. It will be supplied from a branch discharge from the auxiliary feed-pumps in main fire-rooms. It will be fitted with a glass water-gauge, an overflow-pipe, and a float-valve in the auxiliary boiler feed-pump suction-pipe.

AUXILIARY BOILER FEED-PUMP.

There will be in the auxiliary boiler fire-room a steam-pump of approved pattern. It will draw water from the sea and from the auxiliary feed-tank, and will deliver into the auxiliary boiler feed-pipe only.

BLOWERS.

There will be eight blowers of approved pattern, two in each fire-room.

These blowers must be capable of supplying to the fires continuously, with ease, sufficient air to maintain the maximum rate of combustion. They will take air from the fire-room in such manner as to thoroughly ventilate it, and deliver into the air-ducts.

The spindle-bearings are to be accessible while the blowers are in motion, and will be of anti-friction metal in composition boxes, and, together with their lubricating apparatus, must be thoroughly protected from dust.

BLOWING-ENGINES.

Each blower will be driven direct by an inclosed three-cylinder engine of an approved design, and of sufficient power to run the blower at full speed with steam of 100 pounds boiler pressure.

All working parts must be closed in, but easily accessible for overhauling. The lubrication must be automatic and thorough. The throttle-valve in the steam-pipe of each blowing-engine will be arranged to be worked from the fire-room floor, with suitable index to show how much open. The steam-pipe for each blower to connect with auxiliary steam-pipe.

AIR-PRESSURE GAUGES.

An approved gauge will be fitted in each fire-room to show the pressure in each main air-duct.

A portable gauge will also be supplied to each fire-room, with conveniences for connecting it to the furnaces, uptakes and wherever desired to measure the air pressure.

All these gauges are to indicate pressures in "inches of water."

FIRE-TOOL RACKS.

Racks will be fitted in each fire-room in convenient places for holding all necessary fire-tools.

MAIN FEED-PUMPS.

There will be in the after fire-room of the after boiler compartment, and in the forward fire-room of the forward boiler compartment, a vertical double-acting steam-pump of approved design for a main feed-pump. These pumps will each have a water-cylinder of 8 inches diameter and 12 inches stroke, or of equivalent pumping capacity. The valves are to be metallic, of an approved kind. The steam-cylinders must be of sufficient size to work the pumps at the required speed to feed the boiler when under forced draught. Each main feed-pump

will draw water from the sea and from the feed-tank, and will deliver into the main feed-pipe only. If the steam-valve is moved by a supplemental piston the valve must have a positive motion near the end of the main piston-stroke to prevent piston striking cylinder-head. If a supplemental piston is used, its motion must be horizontal.

AUXILIARY FEED-PUMPS.

There will be in each of the two feeding fire-rooms an auxiliary feed-pump, which will be in construction a duplicate of the main feed-pump. It will draw water from the sea, blow-pipes, feed-tank and bottoms of condensers, and will deliver into the auxiliary feed-pipe, fire-main and overboard.

ASH-HOISTS.

Each fire-room ventilator will have vertical guide-strips of iron on the inside, and be fitted with all the necessary gear for hoisting ashes.

An ash-hoisting engine of approved design is to be fitted in each fire-room hatch or such place as may be directed, of sufficient power to hoist 300 pounds from the fire-room floor to the deck in five seconds with steam of 50 pounds pressure.

It will have a reversing gear, to be worked from the upper deck, with approved adjustable safety-gear to prevent overwinding and to stop the engine when the ash-bucket reaches the fire-room floor. Also to be fitted, if required, with an approved brake to control the drum.

The ash-hoists will be fitted with necessary sheaves, whips, and all appliances for handling ash-buckets complete.

Gear will be fitted as directed for hoisting ashes from the auxiliary boiler fire-room.

ASH-DUMPS.

From each ash-hoist, on the upper deck, permanent overhead rails, suitably supported, will lead to the nearest ash-chutes on each side of the ship. Each of these will be fitted

with a traveler of approved design, with all necessary appliances for carrying the ash-buckets. At the top of each ash-chute a dumping-hopper of approved design will be fitted, so arranged as to fold up out of the way when not in use. The ash-buckets are to be balanced dump-buckets, with all necessary gear complete. All the ash-hoisting and dumping-gear is to be such that the buckets will not have to be lifted by hand.

STEAM TUBE-CLEANERS.

A steam tube-cleaner, of approved design, will be fitted in each fire-room. Steam will be taken from the auxiliary steam-pipe. Sufficient length of steam-hose will be provided to easily reach all the tubes.

BILGE AND FIRE-PUMPS.

There will be in each engine-room a double-acting pump of same capacity as the main feed-pumps. The valves will be corrugated discs of phosphor or alluminum-bronze $\frac{1}{16}$ inch thick, with composition guards and with spiral springs. Each pump will have suctions to the sea and engine-room bilge, and will deliver overboard and into the fire-main and the engine-room water-service pipes. These pumps will be secured in forward and after engine-rooms where directed.

DISTILLING APPARATUS.

Two distillers of approved design will be fitted where directed. They will have a combined capacity of 5,000 gallons of potable water per 24 hours, and will be fitted with efficient filters and with the necessary means for conveying the fresh water to the various tanks. The distillers will be made with shells of sheet-brass, flanges and heads of composition, and coils of copper or brass, thoroughly tinned on each side.

The distillers will take steam from the auxiliary boiler by independent stop-valves and pipes, with a branch connection to the auxiliary steam-pipe.

There will be a pump of approved design and sufficient capacity to circulate water through the distillers. It will draw water from a special sea-valve placed where directed. The water after leaving the distillers will lead forward by a proper pipe with connections for flushing the crew's water-closets, and with branches leading to the officers' water-closets. A bye-pass pipe will be provided, so that water may pass to the water-closets when the distillers are shut off.

MAIN STEAM-PIPES.

A $10\frac{3}{4}$-inch steam-pipe will lead from the flange of each boiler stop-valve of boilers in forward compartment, passing through the uptake to the after fire-room of forward boiler compartment, where the two pipes will unite by a breeches-pipe; thence the pipe will be 15 inches internal diameter, leading through the after boiler compartment and the uptake of after boilers; thence to a cross-pipe in forward engine compartment. Where the 15-inch pipe passes through the bulkhead between the boiler compartments it will be fitted with a slip-joint.

In the after boiler compartment a $10\frac{3}{4}$-inch pipe will lead from the stop-valve of each boiler through the uptakes,—one on each side of the 15-inch steam-pipe from forward boilers, and connected with the cross-pipe in forward engine compartment.

Where the pipes pass through the uptakes they will be made of steel or iron, lap-welded, No. 3 B. W. G. for the 15-inch and No. 5 B. W. G. for the $10\frac{3}{4}$-inch pipe.

There will be a steel die-forged flange riveted on each end of each section of steel pipe. The flanges will, inside the uptake, be riveted together; outside the uptake to be secured by through-bolts and nuts. The pipes will incline aft to drain all water from them where they pass through uptakes. A valve of same size as pipe will be fitted to each steam-pipe in the after part of after boiler compartment. The valves to be operated from the level of the main-deck.

A 1-inch spring safety-valve will be fitted to each main steam-pipe between the boiler stop-valve and the main stop-valve.

The cross-pipe in forward engine compartment to be 15 inches internal diameter—made of composition. To the starboard end of this cross-pipe will be bolted a 15-inch straight-way valve to be operated from the main-deck. From this valve a 15-inch pipe will lead to a separator, thence to forward engine throttle-valve.

A 15-inch straight-way valve will be bolted to a 15-inch branch of cross-pipe, pointing aft—this valve to be operated in forward engine compartment.

From this valve a 15-inch pipe will lead to separator of after engine in forward engine compartment, thence to engine throttle-valve of after engine.

A branch from each main steam-pipe, 3 inches in diameter, with proper stop-valve, will lead to the corresponding intermediate valve-chest.

AUXILIARY STEAM-PIPES.

An auxiliary steam-pipe $4\frac{1}{2}$ inches internal diameter will extend through engine and boiler compartments.

Branches will connect in each engine compartment with reversing-engine, air-pumps, circulating-pump and fire-pump in that compartment, and with feed-pumps blowers, and each ash-hoisting engine in each boiler compartment. There will be a connection with the main steam-pipe in each compartment of sufficient size to supply the reversing-engine and the engines of all pumps, blowers and hoists in that compartment.

The auxiliary steam-pipe will also connect with auxiliary boiler. There will be a stop-valve on each side of each watertight bulkhead. Branches will connect with the whistle, the siren and the radiators. Branches will also be led to the engineer's workshop, the dynamo-engines, the ventilating-fan engines, the windlass, the steering-engine and the machinery

in torpedo-rooms; but the connections will be made by those who furnish the machinery in those places. Stop-valves will be fitted wherever necessary. There will be a steam-gauge in brass case, with about 6-inch dial, attached to the auxiliary steam-pipe in each engine-room and each boiler-compartment; also one at windlass and one at steering-engine.

BLEEDER-PIPES.

A 5-inch branch from the main steam-pipe in each engine-room will lead to the condenser, with a stop-valve operated from the working-platform.

AUXILIARY EXHAUST-PIPES.

An auxiliary exhaust-pipe will be fitted, with branches leading to all pumps, blowers and ash-hoists. It will have valves to direct the exhaust-steam into either condenser or into the auxiliary escape-pipe at will. Branches will be led to the dynamo-engines, ventilating-fan engines, windlass, steering-engine, torpedo machinery and workshop engines, but not connected.

Where the auxiliary exhaust-pipe connects with the condenser and with the escape-pipe it will be fitted with two stop-valves at each connection.

MAIN FEED-PUMP EXHAUST.

The exhaust-pipes from the main feed-pumps, in addition to the connection with the exhaust-main, will be so arranged that the exhaust steam can be turned into the feed-pump suction instead of into the auxiliary exhaust-pipe,—chambers with suitable nozzles for this purpose being fitted in the suction-pipes.

If so directed, the exhaust of all pumping-engines will discharge into the feed-pump suctions as provided above for the main feed-pump exhaust.

ESCAPE-PIPES.

There will be an 11-inch escape-pipe abaft each smoke-pipe, extending to top, finished and secured in an improved manner. Each pipe will have branches leading to all the safety-valves in its boiler compartment. There will also be abaft the after smoke-pipe, fitted the same as the others, a 10-inch escape-pipe, connected to the auxiliary exhaust-pipe as above specified. Each of the safety-valve escape-pipes will be fitted with an improved muffler.

MAIN FEED-PIPES.

A pipe will lead from each main feed-pump and discharge only into the main check-valves on the main boilers in the same compartment.

AUXILIARY FEED-PIPES.

A pipe will lead from each auxiliary feed-pump to each auxiliary check-valve on main boilers in its compartment.

BLOW-PIPES.

Pipes will lead from the surface and bottom blow-valves of main boilers in each compartment to a sea-valve in the same compartment. The blow-pipes will have valve and pipe connections to the suctions of the auxiliary feed-pumps in fire-rooms.

A blow-pipe will lead from the auxiliary boiler to the forward fire-room of forward boiler compartment, and be connected with a sea-valve in that compartment.

HOSE CONNECTIONS.

Each bilge and fire-pump and each auxiliary feed-pump will deliver into the fire-main fitted by the contractors for the hull. Standard hose connections, each with screw-cap and a straight-way valve, are to be fitted to the fire delivery-pipes of these pumps; also in each fire-room to a fire-main which

will run from each auxiliary feed-pump to both fire-rooms in the same compartment.

There is to be attached to each of these connections an approved hose of sufficient length to reach to all parts of the compartment in which it is situated, including the connecting coal-bunkers. Each hose is to be fitted complete with standard brass couplings, nozzles and spanners, and to have approved means of stowing while attached to its fire-plug.

ENGINE-ROOM WATER-SERVICE.

There will be in each engine-room a 3-inch pipe connected with the sea-suction valve and with the bilge and fire-pump delivery, with branches leading to the different parts of the engine, as follows :

One $1\frac{1}{2}$-inch pipe to the back of each crank-shaft bearing ;

Two $1\frac{1}{4}$-inch pipes, fitted with detachable spray-pipes, as directed, to each crank-pin ;

Two 1-inch pipes, fitted with detachable spray-pipes, as directed, to each cross-head ;

One $1\frac{1}{4}$-inch pipe to the water-jacket of each cross-head guide ;

Two 1-inch pipes to each air-pump engine ;

Two 1-inch pipes to each circulating-pump engine ;

From the after engine-room, in addition to the pipes above specified, there will be a 2-inch pipe, with branches, as may be directed, to each thrust-bearing ; also a 1-inch pipe to each line-shaft bearing.

Each branch will have a separate valve.

All the water-service pipes and fittings above the floors are to be of polished brass.

Where directed the pipes are to have pivoted nozzles.

The water-service pipes of the two engine-rooms are to be connected with each other by a 3-inch pipe and valve.

SEA-SUCTION PIPES.

A pipe will lead from the sea-suction valve in each engine-

room to the bilge and fire-pump in its compartment. In each boiler compartment pipes will lead from the sea-suction valves to the main and auxiliary feed-pumps, and in the forward boiler compartment to the auxiliary boiler feed-pump. A pipe will also lead from a sea-valve, fitted where directed, to the distiller circulating pump. Each of these pipes will be of at least the same bore as the nozzle on the pump with which it connects. Each sea-suction will be controlled by a valve which will not permit sea-water to enter any of the bilge suction-pipes or feed-tank suction-pipes.

FEED-TANK SUCTION-PIPES.

From the feed-tank a suction-pipe will lead to the main and auxiliary feed-pumps in the after boiler compartment. A similar pipe will lead from feed-tank to the pumps in the forward boiler compartment. Each pipe to have valves at tank and pumps.

FEED-SUCTIONS FROM CONDENSERS.

From the bottom of each condenser a pipe will lead forward, the two pipes joining in the forward engine-room; thence branches will lead to both suction-pipes from feed-tank. There will be a screw non-return valve in each pipe close to condenser.

SUCTIONS FROM DOUBLE-BOTTOM VALVE-BOXES.

Suction-pipes will lead from the double-bottom valve-boxes to the auxiliary pumps as directed.

These pipes will be connected also to the forward engine-room sea-suction pipes for the purpose of filling the double-bottom.

BILGE-SUCTION PIPES.

Each bilge and fire-pump will be connected with the engine-room bilges and the drainage system of the ship,—each pipe having a non-return valve and a Macomb, or equivalent, bilge-strainer above the floors.

BOILER PUMPING-OUT PIPE.

A suction-pipe, leading to the auxiliary feed-pumps, will connect with the main blow-pipe in each boiler compartment.

PIPE CLOTHING.

All steam and exhaust-pipes, including their flanges, the separators and all steam-valves are to be clothed in an approved manner with a satisfactory incombustible non-conducting material, covered with canvas in double thickness, well painted. The canvas covering of steam-pipes to be secured to bulkheads where the pipes pass through them. The main steam-pipes in engine-rooms and the main exhaust-pipes and the separators are also to be covered with black-walnut lagging, in sections, with brass bands.

SEPARATORS.

There will be in each main steam-pipe, in the forward engine compartment, a separator of approved design. Each will be fitted with a well-protected glass gauge, also with an approved automatic steam-trap with a 2-inch drain delivering into feed-tank; also with a 5-inch drain-pipe leading overboard. Valves to be placed in drain-pipe as directed.

PIPES THROUGH COAL-BUNKERS.

All pipes passing through coal-bunkers will be protected by sheet-iron casings, made in sections and easily removable for repairs. Pipes must not be led under openings of coal-chutes.

PIPES THROUGH BULKHEADS AND DECKS.

All pipes passing through water-tight bulkheads or decks will be made water-tight by stuffing-boxes, flanges or other approved means. Pipes must not be led in such a way that the angles or tees of bulkheads have to be cut. Holes through wooden decks, where pipes pass through, are to have brass or copper thimbles, made water-tight, and extending at least 3 inches above the deck.

MATERIAL AND FITTING OF PIPES.

All pipes, except the lower ends of bilge-suction pipes, or where otherwise specified, will be of copper. The lower parts of bilge-suction pipes are to be made of galvanized iron.

All feed and blow-pipes, all bilge-suction pipes, except the lower parts, and all steam pipes less than three inches diameter, are to be seamless-drawn. All copper pipes not seamless-drawn are to be brazed.

All copper pipes over three inches diameter will have composition flanges riveted on and brazed; under three inches to have flanges or approved composition couplings brazed on. All feed and blow-pipes to have composition flanges.

All flanges are to be faced and grooved, and joints made with approved material.

All composition flanges below the floor-plates are to be connected by naval-brass bolts and nuts.

All copper pipes in bilges are to be well painted and covered with waterproof canvas, and must not rest in contact with any of the iron or steel work of the vessel. All bends in copper pipes are to be made one gauge thicker than straight parts. All copper pipe fittings are to be made of composition. Expansion joints of approved pattern are to be fitted where required.

Slip-joints are to have stop-bolts and flanges where directed.

THICKNESS OF PIPES.

The thickness of metal in the principal pipes will be as follows by B. W. G.:

Steam-pipes of 15 inches bore _____ No. 3.
Steam-pipes of $10\frac{3}{4}$ inches bore _____ No. 5.
Steam-pipes of 5 inches bore _____ No. 9.
Steam-pipes of 4 inches bore _____ No. 10.
Steam-pipes of 3 inches bore _____ No. 11.
H. P. exhaust to I. P. cylinder _____ No. 7.

I. P. exhaust to L. P. cylinder_____No. 7.
L. P. exhaust to condenser_____No. 7.
Circulating-pump suction and discharge-pipe_____No. 5.
Bilge-injection pipes _____No. 11.
Air-pump discharge to feed-tank_____No. 11.
Feed-pump suction-pipes_____No. 13.
Feed-pump discharge-pipes_____No. 10.
Blow-pipes _____No. 9.
Auxiliary exhaust-pipes_____No. 14.
Escape-pipes_____No. 13.
Dry-pipes _____No. 14.
Connections to fire-main_____No. 10.
Galvanized wrought-iron bilge-suction pipes_____No. 7.

All other pipes to be of such thickness as may be approved.

TURNING-ENGINES AND GEAR.

There will be in each engine-room a double 6 x 6-inch engine, driving, by worm-gearing, a second worm, which may be made at will to mesh with a worm-wheel on the propelling-shaft. The worm-wheel of the forward engine is to be fitted on the flange-couplings at after end of crank-shaft; that of the after engine is to be bolted to the forward coupling-flange of crank-shaft.

The turning-engine shaft will be squared at the end and fitted with a ratchet-wrench, of approved design, for turning by hand.

Each turning-engine will have piston-valves, and will be made reversible by means of a change-valve moved by a screw and hand-wheel.

The turning-wheels will be of cast-steel with cut teeth.

DRAIN-PIPES AND TRAPS.

All places where condensed steam can accumulate will be provided with drain-pipes and cocks or valves of ample size, and with approved automatic traps, which will discharge into

feed-tanks or condensers or as directed. All traps are to have bye-pass pipes and valves for convenience of overhauling. The lowest parts of all water-pipes and all pump-cylinders and channel-ways are to have drain-cocks, with pipes where required.

The handles of all drain-cocks will point downward when closed.

AUXILIARY ENGINE STOP-VALVES.

Each auxiliary engine will have stop-valves in both steam and exhaust-pipes as close as possible to the cylinders. Exhaust stop-valves are to be straight-way where practicable.

All pumps, except circulating-pumps, will have screw check-valves in both suction and delivery-pipes close to pump-cylinders, so arranged that they may be kept off their seats when desired.

PUMP RELIEF-VALVES.

All feed and fire-pumps will have adjustable spring relief-valves of approved design, connecting the delivery and suction passages.

SEA-VALVES.

There will be in the various compartments sea-valves as follows:

In each engine-room a 5-inch sea-suction valve with a 5-inch nozzle for pump-suction, and a 3-inch nozzle for engine-room water-service; to be a screw stop-valve and attached to ship where directed;

A double valve-box with two screw non-return valves, one 5 inches for main bilge-pump discharge and one 4 inches for bilge and fire-pump discharge, from which a composition pipe will lead to side of ship; also to have a 2-inch non-return valve attached for trap discharge;

A main injection-valve and a main outboard-delivery valve as before specified.

In each boiler compartment a 5-inch sea-suction valve with two 5-inch nozzles for feed-pump connections;

A 3-inch valve, connected to the blow-pipes by a 3-inch nozzle.

There will be also a sea-suction valve for the distiller circulating-pump, placed where directed. These valves to be screw stop-valves, connected to bottom of ship through the double bottom.

All sea-suction valves are to have gratings on the outside of the ship. No waste water is to be delivered above the water-line.

ATTACHMENT OF VALVES TO HULL.

Steel strengthening-rings will be riveted to plating of hull around the openings for all sea-valves. The valve-flanges will be bolted to these rings by naval-brass studs, care being taken not to drill the holes entirely through the rings. Attachments will be made to the inner skin as shown in drawings. A zinc protecting-ring will be fitted in each opening in outer skin and at the openings in inner skin for the main injection and outboard-delivery valves.

BILGE AND DOUBLE-BOTTOM VALVE-BOXES.

The bilge and double-bottom suction valve-boxes and valves are to be supplied by the engine contractors, who will also connect them with the pumps, but the connections to them from the bilge-drains and from the forward and after compartments will be made by the hull contractors. There is to be a separate valve for each suction, and they are all to be plainly marked to show their uses.

COCKS AND VALVES.

All cocks and valves and their fittings, unless otherwise specified, are to be made of composition. All hand-wheels are to be of finished brass except as otherwise specified, and will be at least one and a half times as great in diameter as their valves.

All cocks communicating with vacuum spaces are to have bottoms of shells cast in and to have packed plugs. All cocks over one inch in diameter are to have packed plugs. Valves of approved pattern are to be supplied wherever necessary to complete the various pipe systems, whether herein specified or not. Cocks and valves, where directed, have, in lieu of wheels or permanent handles, removable box or socket-wrenches, marked and stowed in convenient racks. All valve-spindles are to turn right-handed to close.

All cocks and valves beneath the floor-plates are to have their wheels or handles above the floor-plates, except where otherwise specified, in easily accessible positions. All valves must be so fitted as to be easily ground in, and be fitted where necessary with grinding-in guides and handles.

No conical-faced valve must have a bearing on its seat of more than $\frac{3}{16}$ inch in width.

The size of valves as given in these specifications refers to the diameters of the equivalent clear openings.

LABELS ON GEAR AND INSTRUMENTS.

All cocks are to have engraved brass plates to show their uses and to indicate whether open or shut. All valves, except such as may be otherwise directed, are to have similarly engraved plates to show their uses, or have the same plainly engraved on hand-wheels.

All hand-levers will be similarly marked. Gear for working valves from deck will be marked as elsewhere specified.

All main steam stop-valves are to have indices to show to what extent they are opened.

All gauges, thermometers, counters, telegraph-dials, speaking-tube annunciators and revolution-indicators will be suitably engraved to show to what they are connected.

All engraving is to be deep and to be filled in with black cement.

WHISTLE.

An approved polished brass steam-whistle, with bell of

about 8 inches diameter, is to be placed forward of the forward smoke-pipe, well above the level of the awnings, and connected to the auxiliary steam-pipe by a pipe having a stop-valve at its lower end and a working-valve at the upper end; the pipe to have an expansion-joint at lower end.

SIREN.

There is to be a steam-siren of approved pattern and size, placed where directed, and connected similarly to the whistle.

RADIATORS.

Radiators or heating-coils, as may be directed, of approved patterns, will be furnished, fitted and connected, with superficial areas as follows:

In the commanding officer's cabin, four of 20 square feet each;
In the commanding officer's bath-room, one of 2 square feet;
In the commanding officer's office, one of 2 square feet;
In the ward-room, four of 20 square feet each;
In the steerage, two of 15 square feet each;
In the steerage country, two of 15 square feet each;
In the crew's quarters on the berth-deck, an aggregate of 350 square feet, divided as may be directed;
In the sick bay, two of 15 square feet each;
Under the forecastle, two of 20 square feet each;
In the pilot-house, one of 6 square feet;
In the chart-house, one of 4 square feet.

Each radiator or coil of more than 6 square feet surface is to be divided into two parts, and each of more than 20 square feet into three or more parts—each with its separate steam and drain-valves. The radiators in the crew's quarters will have the valve-stems squared and fitted with removable keys. The steam and drain-pipes are to be of seamless-drawn brass, iron pipe size, suitably connected by composition fittings in a manner that will enable them to be easily taken down for

repairs. All union-joints to be coned or to have corrugated copper washers. All holes through decks and bulkheads to be thimbled with brass. Steam and drain-pipes to be clothed where near wood-work. The steam-pipes will connect with the auxiliary steam-pipe where directed, and will be fitted with adjustable reducing-valves. The drain-pipe of each circuit will have an approved steam-trap discharging into feed-tank, and elsewhere as may be directed.

SHAFTS THROUGH BULKHEADS.

All shafts passing through water-tight bulkheads will be fitted with stuffing-boxes.

FLOOR-PLATES.

The engine-rooms, fire-rooms, and connecting passages are to be floored with wrought-iron plates $\frac{1}{4}$ inch thick, with neatly matched flat-topped corrugations running fore and aft. The plates are to be of convenient size and easily removable, except over air-ducts in fire-rooms. They will rest on proper ledges of angle or T-iron, and will have drain-holes where necessary. Platforms will be provided for getting at all parts of the main and auxiliary engines and boilers. These platforms, where placed over moving machinery, will be fitted the same as the lower floors. In other places they will be made of iron-rods $\frac{5}{8}$ inch square, placed $\frac{3}{4}$ inch apart. The fire-room floors will be so arranged as to permit all furnaces to be easily fired without interfering with getting out coal.

LADDERS.

Ladders will be fitted wherever necessary for reaching the engine-rooms and fire-rooms from deck, and for reaching the various platforms, passages, and parts of machinery. The engine-room ladders will be made with plate-iron sides and light cast-iron treads with corrugated tops. The fire-room ladders will be made with plate sides and double square-bar treads.

All ladders will be so fitted as to be easily removable where required, and will be jointed and hinged, with necessary fastenings and gear, where they have to be moved when closing hatches. Light iron ladders will be fitted to and through one ventilator in each engine-room and each fire-room as means of egress when the battle-hatches are closed, those in the fire-room ventilators to be removable.

HAND-RAILS.

Finished brass hand-rails with finished wrought-iron stanchions, easily removable where required, will be fitted to all ladders and platforms, and around all moving parts of machinery, and along bulkheads and passage-ways.

The lower ends of stanchions to pass through floor-plates, with nuts underneath.

WORKING-PLATFORMS.

There is to be a working-platform between each high-pressure and intermediate engine, over the engine-frame. The counter, revolution-indicators, clock, gauges, telegraph-dials and other engine-room fittings are to be so placed near the working-platform as to be in full view while working the engines. Speaking-tube mouth-pieces and telegraph-levers to be conveniently placed.

WORKING-LEVERS AND GEAR.

There will be at each working-platform the following hand-gear, viz:

One reversing-lever;
Three starting-valve levers;
Three cylinder drain-cock levers;
Hand reversing-wheel;
Throttle-valve hand-wheel;
Bleeder-valve hand-wheel.
Reversing-engine stop-valve hand-wheel;
Starting-valve stop-valve hand-wheel.

All levers are to have spring-catches of "locomotive pattern."

ENGINE-ROOM INSTRUMENTS.

Each engine-room will be fitted with the following, in full view of the working-platform and properly lighted, viz:

One Lane's improved spring steam-gauge, connected to main steam-pipe;

One Lane's improved spring steam-gauge, connected to the high-pressure exhaust-pipe;

One Lane's improved spring compound-gauge, connected to the intermediate exhaust-pipe;

One Lane's improved spring compound-gauge, connected to the main condenser;

One eight-day clock with second-hand;

One continuous rotary counter with positive motion, to register from 1 to 1,000,000;

Two revolution-indicators, showing on suitable dials the speed of both screw-engines;

One telegraph-dial.

There will also be in each engine-room the following instruments, attached directly to the various parts of the machinery:

One hot-well thermometer, to be connected to the hot-well discharge-pipe;

One injection-water thermometer;

One discharge-water thermometer;

One steam-thermometer in main steam-pipe;

One mercurial vacuum-gauge on main condenser;

One counter or revolution-indicator on the main circulating-pump;

One continuous rotary counter for air-pump with positive motion; to register from 1 to 1,000,000.

Each working-platform will be fitted with telegraph and speaking-tubes leading to engine-rooms, fire-rooms, wheel-house, conning-tower, bridge and where directed.

Each engine-room will be furnished with eight approved

indicators of standard size of the latest pattern, with barrels of sufficient size to make diagrams not less than 5 inches long; two to have, each, a spring of 80 pounds to the inch and one of 64 pounds; two to have, each, a spring of 40 pounds to the inch and one of 32 pounds; each of the other four indicators to have a spring of 20 pounds to the inch and one of 16 pounds. All indicators to be nickel-plated and to be complete with all attachments. One extra cock-attachment to be furnished with each indicator. Each indicator to be in its own case and to be suitably marked. Cases to be conveniently stowed.

There will also be in each engine-room a steam-gauge on the auxiliary steam-pipe, as before specified; also steam-gauges on the steam-pipes to intermediate and low-pressure cylinder-jackets. All instrument casings and fittings to be of polished brass. All thermometers to be permanent fixtures, with stems and bulbs protected.

Each blowing-engine in fire-rooms will be fitted with an approved revolution-indicator. All gauges, thermometers, and revolution-indicators in engine-rooms will be suitably engraved, on dials or elsewhere, to show to what they are connected.

LOG-DESKS.

A desk of approved pattern, with locked drawer, is to be fitted in each engine-room where directed.

INDICATOR FITTINGS AND MOTIONS.

An indicator connection is to be made to each end of each cylinder of main engines, and to each end of each air-pump as near as possible to the bore of the cylinder, and so as to be easily accessible. Each indicator when in place is to be connected to but one end of a cylinder. The connecting-pipes are to be one-inch bore, without bends. The indicator motion of each engine and air-pump is to be so fitted that both indicators on its cylinder can be connected at the same

time. The motions of the indicator-barrels must be accurately coincident with the motion of the corresponding pistons, and such as to give a motion of not less than 5 inches. The steam-cylinders of all auxiliary engines are to have holes tapped for indicator fittings and then plugged. These engines are to have portable indicator motions fitted, then removed and suitably marked and stowed. When auxiliary engines are duplicated but one set of indicator-motion fittings need be supplied for all of each kind.

ENGINE-ROOM TELEGRAPHS.

A repeating telegraph of approved pattern is to be fitted in each engine-room and connected to transmitters in conning-tower, in wheel-house and on bridge. The connections are to be made in such manner that the chance of derangement shall be minimized.

SPEAKING-TUBES.

Speaking-tubes will be made of copper not less than No. 20 B. W. G. They will connect each engine-room with each fire-room; the engine-rooms with each other; the fire-rooms with each other; each engine-room to the wheel-house, conning-tower, bridge and to the chief engineer's room; each fire-room with the upper deck close to the top of the ash-hoist, and elsewhere as required. Each tube will be fitted at each end with a mouth-piece and approved annunciator; the mouth-pieces to be connected to short flexible pipes where required. All mouth-pieces or pipes to be plainly marked. The tubes will be suitably cased where necessary.

REVOLUTION-INDICATORS.

Revolution-indicators are to be of such approved pattern as shall not be affected to a perceptible degree by the motion of the ship or by changes of temperature. They must be worked off the engines by positive motions, and must be so fitted that changes of engine speed shall not produce violent fluctuations of the indices.

TELL-TALES.

Tell-tales with proper connections are to be fitted on the bridge and in the conning-tower, to show the direction of motion of the propelling-engines.

LUBRICATION.

All working parts of machinery are to be fitted with efficient lubricators, with capacity for sufficient oil for four hours' running. Each main crank-pin is to have a centrifugal oiler each to be fed by a pipe from an oil-box on top of the adjoining main bearing, also a telescopic oiling-gear. Each main bearing is to have four sight-feed oil-cups with well-protected glass tubes, to be of approved pattern. The cross-heads will each have a wiping oiling-gear fitted at the outboard end of the stroke. The thrust-bearings will have oil-boxes as before specified. There will be on each valve-chest cover a globe oil-cup to lubricate the valve-stems; also a globe oil-cup over each valve-port. Each eccentric will have a long oil-cup, fed by a drip-pipe, so arranged that the eccentric will be thoroughly lubricated in all positions. There will be fitted to each main steam-pipe, close to valve-chest, a Siebert or equivalent steam sight-feed oil-cup of two quarts capacity. There will be smaller steam sight-feed cups on each circulating, blowing, main-feed and bilge-pump engine; all steam sight-feed cups to have ample condensing surface on their steam-pipes. Each three-cylinder engine will have a continuous automatic lubricator of approved pattern. All working parts for which oil-cups are not specified or shown in drawings will have oiling-gear of approved design, such that they can be oiled without slowing. All the oiling of each auxiliary engine to be done by one oil-box where practicable. All fixed oil-cups are to have hinged covers, with stops to prevent being opened too far. Moving oil-cups, where necessary, will have removable covers. The supply of oil to various parts is to be easily regulated. All oil-cups and their fittings, except such as are cast

on bearings, are to be of finished cast-brass, or of sheet-brass or copper, as may be directed, with all seams brazed.

OIL-DRIPS.

All fixed bearings will have drip-cups cast on where possible, otherwise they will be of cast-brass, properly applied. All moving parts will have drip-cups or pans cast on engine-frames where directed, otherwise to be substantially made of sheet-brass or copper with brazed seams. All drip-cups will have drain-pipes and cocks of at least $\frac{1}{2}$ inch diameter which can be used while the engines are in operation.

ASH-SPRINKLERS.

A valve for wetting down ashes will be fitted in each fire-room where directed, and will be fitted with all necessary hose, couplings and nozzles.

JOURNAL-BOXES.

All journals or moving parts of iron or steel are to run, unless otherwise specified, in composition boxes. These boxes are to be lined with approved anti-friction metal where directed.

MANDRELS FOR WHITE-METAL BEARINGS.

Hollow cast-iron mandrels, as shown, are to be furnished for forming the white-metal linings of crank-pin, crank-shaft, line-shaft, and thrust-bearings. All these to be smoothly and accurately turned to size and packed so as to be perfectly protected.

STUFFING-BOXES.

All iron stuffing-boxes are to have composition bushings. All glands are to be of composition. Metallic packing will be fitted, as elsewhere specified. All glands not fitted with pinion-nuts and spur-rings will have lock-nuts and split-pins.

PUMP-CYLINDERS.

All pump-cylinders, together with their valve-boxes and fittings, will be made of composition. Air-chambers will be fitted on the delivery sides of pumps or in the pipes, as may be directed.

BOLTS AND NUTS.

All bolt-heads and nuts, except in special cases, are to conform to the United States Navy standard. Screw-threads on bolts and nuts less than 2 inches must in all cases conform to the above standard. All finished bolts, except as directed, are to be kept from turning by dowels or other suitable devices.

The nuts of all bolts on moving parts and on pillow-blocks, and elsewhere as shown, are to be fitted with keepers, and the bolts will extend beyond the nuts, without threads, and fitted with split-pins. The nuts on piston-rods will have screw stop-pins.

GEAR FOR WORKING VALVES FROM DECK.

The rods for working valves from the main deck will be guided and supported on deck by cast composition standards, left rough and painted. Each rod will have a hand-wheel at least 4 feet above the deck. The stop-valve hand-wheels are to be 16 inches in diameter; each to be fitted with an approved lock and key. The wheels will be of brass, polished, and will have their rims connected with the hubs by plain discs without holes in them. Or in lieu of hand-wheels, if directed, polished brass bar-handles will be fitted to squares on the turning-rods, and will be stowed in beckets on bulkheads. All hand-wheels to be engraved with name; or cast-brass label-plates with polished raised letters will be fixed to adjoining bulkheads.

SUPPORTS FOR LAMPS.

Supports for lamps will be fitted as and where directed.

LIFTING-GEAR.

Efficient lifting-gear, consisting of traveler-bars and pulleys, deck-beam clamps, turn-buckles, shackles, hooks, eye-bolts, and as may be directed, will be fitted wherever required for lifting parts of the machinery for overhauling and repairing.

OIL-TANKS.

Four oil-tanks, of 250 gallons capacity each, will be fitted where directed, with facilities for filling from deck. They will be made of wrought-iron not less than $\frac{1}{8}$ inch thick, and each will have a man-hole and cover near the top and a locked cock for drawing oil. In each engine-room there will be fitted two copper oil-tanks of 20 gallons each and two of 8 gallons each, and in each boiler compartment one of 5 gallons, all with lock-cocks. All oil-tanks will be fitted with drip-pans.

An iron tallow-tank, divided into two parts, with hinged covers, will be fitted where directed.

VENTILATORS.

Ventilators with cowls well above the awnings are to be fitted, two 27-inch to each fire-room, and three to each engine-room, 36 inches diameter above upper-deck and 30 by 42 inches below.

Each fire-room ventilator will be fitted as an ash-hoist and provided with all the necessary gear for hoisting ashes, and with door at the side of the ventilator at the upper deck to remove the ash-buckets. The ventilator-shafts are to be made of wrought-iron or steel not less than $\frac{1}{8}$ inch thick, and the cowls will be made of copper, No. 12 B. W. G., not planished. The base-rings of cowls will be of cast composition, finished on working parts, but left unfinished on the outside. All cowls are to be fitted with gear for turning them from engine-rooms and fire-rooms; the gear to be of composition, except the spindles, which will be of wrought-iron. Brass hand-wheels or T-handles will be fitted to spindles in engine and fire-rooms.

INSTRUMENTS.

The following instruments and tools are to be furnished in addition to those elsewhere specified, viz:

Two indicators of the same kind as furnished for the main engines; each with springs of 80, 64, 40 and 20 pounds to the inch; each in its own case, with fittings complete, and with two extra cock attachments;

Two spare water-thermometers;

Two spare steam-thermometers;

One standardized thermometer in suitable case, with certificate of standardization;

A fixed trammel for setting the main slide-valves without removing covers; the valve-stems to be properly marked for this purpose;

Fixed trammels or gauges for aligning crank-shafts; brass pins to be let into bases of pillow-blocks and into engine-frames and center-marked for this purpose.

All trammels and gauges to have protecting cases.

TOOLS.

One set of wrenches complete for each engine-room and for each boiler compartment, to be fitted to all nuts in their respective compartments, plainly marked with sizes, and fitted in iron racks of approved pattern. The wrenches for nuts of bolts less than 1 inch diameter will be finished, and for all over 2 inches diameter will be box-wrenches, where such can be used. Socket-wrenches to be furnished where required. Open-end wrenches are to be of steel or of wrought-iron, with case-hardened jaws; all others to be of wrought-iron or cast-steel.

One pair of taps, on rod, for tapping front and back tube-sheets of main boilers at one operation. This is to be a duplicate of that used in originally tapping the sheets, and is to be so packed as to be perfectly protected from injury.

A similar pair of taps for the auxiliary boilers.

Two complete sets of fire-tools for each fire-room.

Twelve coal-buckets and twelve ash-buckets.
All tools to be conveniently stowed.

DUPLICATE PIECES.

The following duplicate pieces, in addition to others specified, are to be furnished, fitted and ready for use, viz:

One set of valves for each pump;

One valve-seat, with guards and bolts complete, for air-pumps;

One-half set of follower-bolts and nuts for each steam-piston, and one set for each air-pump piston;

One-half set of springs for each steam-piston;

Two back brasses and two front brasses for crank-shaft journals;

Two crown brasses and two butt brasses for crank-pin journals;

Two crown brasses and two butt brasses for cross-head journals;

Two composition shoes for cross-heads;

One section of crank-shaft, to be fitted in place, and delivered at such naval station in the United States as may be directed, to be left in store;

One complete set of brasses for each main engine valve-gear;

One complete set of brasses for each circulating-pump engine, each air-pump engine, each main feed-pump, each fire-pump and each blowing-engine;

One hundred stay-tubes for main boilers, threaded to fit threads in tube-sheets, with ends wrapped in canvas;

One hundred ordinary boiler-tubes for main boilers, swelled at one end and annealed, ready for use;

One complete set of stay-tubes for auxiliary boiler, ready for use;

All boiler-tubes to be securely stowed in racks, or as directed;

Two hundred condenser-tubes, packed in boxes;

Fifty condenser-tube glands;

One spare spring for each safety-valve and relief-valve;
One spare basket for each Macomb bilge-strainer;
One-eighth of a set of grate-bars and bearers for all furnaces.

All duplicate pieces, not of brass, except as otherwise specified, are to be painted with three coats of white lead and oil and well lashed in tarred canvas, with the name painted on outside. Brass pieces are to be marked or stamped. All pieces to be stowed in an approved manner.

SECURING ENGINES IN SHIP.

The bases of crank-shaft pillow-blocks will be secured to engine-seatings by body-bound through holding-down bolts, with locked nuts at upper ends. Holes in engine-seating and bases to be reamed when the bases are in place. The keys between bases and crank-shaft pillow-blocks to be fitted out of wind.

The inboard foot of each cylinder to rest on wrought-iron fitting pieces about $\frac{3}{4}$ inch thick; these pieces to be riveted to the engine-seating and surfaced perfectly parallel to each other and to the upper surfaces of keys in bases of crank-shaft pillow-blocks. The cylinders to be securely bolted to these pieces—iron to iron—by body-bound through holding-down bolts, with locked nuts at upper ends; the bodies of bolts to be turned, and the holes in seatings and cylinder feet to be reamed in place. The cylinder axes must each be exactly perpendicular to the axis of the crank-shaft and pass through the middle of the length of the crank-pin. The outer supports of cylinders and the inboard supports of cross-head slides to rest on iron washers, and be secured to the seatings by finished through-bolts having locked nuts.

The ends of cross-head slides next the cylinders to be secured to the cylinders as shown in drawings.

The line-shaft and thrust-bearing pillow-blocks to be secured to the vessel in same manner as supports for cross-head slides.

When finally secured all shafting must be accurately in line with the vessel at load-draught and ordinary stowage.

The air-pumps, circulating-pumps and all auxiliary engines to be secured to their seatings in the same manner as supports for cross-head slides; all parts of machinery and boilers are to be secured in an approved manner to prevent displacement when the vessel is used for ramming. The contractors for the hull will supply the labor to fit the seatings to the engines, boilers and auxiliaries.

MATERIALS AND WORKMANSHIP.

All castings must be sound and true to form, and before being painted must be well cleaned of sand and scale and all fins and roughness removed.

No imperfect casting or unsound forging will be used if the defect affects the strength or to a marked degree its sightliness.

All nuts on rough castings are to fit facings raised above the surface. All flanges of castings are to be faced, and those coupled together are to have their edges made fair with each other. The facings of all circular flanges are to be grooved.

All bolt-holes in permanently fixed parts are to be reamed or drilled fair and true in place, and the bodies of bolts finished to fit them snugly.

All threads on bolts and nuts less than two inches are to be of the United States Navy standard.

All pipes beneath floor-plates are to be connected by forged bolts and nuts of naval brass.

All brasses are to fit loosely between collars of shafting.

All brasses or journals are to be properly channeled for the distribution of oil.

Packing for stuffing-boxes to be such as may be approved.

All small pins of working parts are to be well case-hardened.

All materials used in the construction of the machinery are to be of the best quality. The iron castings are to be made of the best pig-iron, not scrap.

Composition castings will be made of new materials. The various compositions will be by weight as follows, viz:

For all journal-boxes and guide-gibs, where not otherwise specified:
>Copper, 6 parts;
>Tin, 1 part;
>Zinc, ¼ part.

Naval brass:
>Copper, 62 per cent.;
>Tin, 1 per cent.;
>Zinc, 37 per cent.

For composition not otherwise specified:
>Copper, 88 per cent.;
>Tin, 10 per cent.;
>Zinc, 2 per cent.

Muntz metal to be of the best commercial quality.

Anti-friction metal to be of approved kind.

Ornamental brass fittings are to be of good uniform color.

All castings are to be increased in thickness around core-holes. Core-holes are to be tapped and the core-plugs screwed in and locked.

All steel forgings are to be without welds and to be free from laminations.

All flanges, collars and off-sets to have well-rounded fillets.

All boiler-plates are to be well cleaned of oxide-of-iron scale.

All flanged parts of boilers are to be annealed, after flanging, in an approved manner.

All bolts for securing the boiler attachments are, where possible, to be screwed through the boiler-plates with heads inside.

All work is to be in every respect of the first quality and executed in a workmanlike and substantial manner.

Any portion of the work, whether partially or entirely completed, found defective, must be removed and satisfactorily replaced without extra charge.

TESTS OF MATERIALS.

All materials used in construction of the boilers, shafting and steel castings will be tested in accordance with the "Instructions to Inspectors" accompanying these specifications.

All boiler and condenser-tubes must be tested to 300 pounds internal water-pressure before being put in place. India-rubber valves taken at random must stand a dry-heat test of 270° F. for one hour and a moist heat test of 320° for three hours without the quality being impaired.

TESTS OF BOILERS AND MACHINERY.

Before the boilers are painted or placed in the vessel they will be tested under a pressure of 225 pounds to the square inch above atmospheric pressure. This pressure is to be obtained by the application of heat to water within the boilers, the water filling the boilers quite full.

Each high-pressure cylinder, jacket, and valve-chest, the steam-pipes and valves, the auxiliary engines, and all fittings and connections subjected to the boiler pressure are to be tested by water-pressure to 225 pounds to the square inch; the intermediate cylinders and connections to 150 pounds; the low-pressure cylinders and connections to 135 pounds; the condensers to 30 pounds. The pumps, valve-boxes and air-vessels of the feed, fire and bilge-pumps are to be tested to 250 pounds per square inch. The cylinders and condensers are to be tested before being placed on board, and must be so placed that all parts may be accessible for examination by the Inspector during the test. All parts are also to be tested after being secured on board. No lagging or covering is to be on the cylinders or condensers during the tests.

PAINTING.

After a satisfactory test the boilers are to be painted on the outside with two coats of brown zinc and oil, and when in place the fronts will be painted with one coat of black paint.

All engine-work, not finished, to be primed with two coats of brown zinc and oil, and when placed in position on board the vessel will be painted with two coats of paint of approved color. The shafting, when in place, is to be painted with two coats of red lead and oil and two coats of black paint.

The smoke-pipes are to be thoroughly painted before and after erection on board. The ventilators and cowls will be painted similarly to the smoke-pipes, except the interiors of the cowls, which will be painted vermilion.

All steam-pipes not lagged with wood will be painted white; exhaust-pipes, green; water-supply pipes, red; and water-discharge pipes, lead color.

PRELIMINARY TESTS AND TRIALS.

Steam is not to be raised in the boilers until after the water-test on board, unless desired for drying joints, for which purpose the pressure must not exceed 10 pounds per square inch.

After testing, steam will be raised in the boilers whenever required to test the connections and the workings of all parts of main and auxiliary engines. All expense of such preliminary tests will be borne by the contractor.

SUPERINTENDING NAVAL ENGINEER'S OFFICE.

A suitable office and draughting-room, properly furnished, fitted, and heated, is to be furnished by the contractor for the use of the Superintending Naval Engineer and his assistants during the building of the machinery and its erection on board.

RECORD OF WEIGHTS.

The actual finished weights of all machinery and appurtenances thereto as fitted, also, all stores, spare machinery, and tools, will be weighed by the contractor in the presence of the Superintending Naval Engineer, or one of his assistants, who will certify to the same, before being placed in the vessel, and the weights, in detail, will be forwarded once a

month to the Bureau of Steam Engineering; also a statement of the aggregate weights to the Bureau of Construction and Repair.

WORKING-DRAWINGS.

All drawings necessary for the prosecution of the work must be prepared by and at the expense of the contractor. Those which are developments of the drawings furnished and of these specifications will be subject to the approval of the Superintending Naval Engineer before the work is ordered or commenced.

In the drawings furnished figured dimensions, where given, will be followed, and not scale dimensions, unless otherwise directed. All discrepancies discovered in drawings, or between drawings and specifications, will be referred to the Superintending Naval Engineer, who will keep a record of the same, together with his decisions, and report the same to the Bureau of Steam Engineering.

CHANGES IN PLANS.

The contractor must make no changes from the drawings furnished, except as herein directed, or from the provisions of these specifications, without preparing complete plans of such proposed changes and submitting the same to the Superintending Naval Engineer for the approval of the Navy Department.

DRAWINGS OF COMPLETED MACHINERY.

Draughtsmen will be employed by the Bureau of Steam Engineering, who will, under the direction of the Superintending Naval Engineer, make a complete set of general and detail drawings of the machinery and boilers as fitted. These plans will be forwarded to the Bureau of Steam Engineering as the work progresses.

A copy of these drawings will be made by these draughtsmen, to be a part of the outfit of the vessel.

OMISSIONS.

The engines, boilers, uptakes and smoke-pipes, all auxiliaries, piping and connections, all sea-valves (except the cutting of the holes for the same), and all parts described in these specifications and in the official drawings are to be fitted complete to the vessel by the engine contractors; and any part of the machinery, or any article pertaining thereto, which may have been inadvertently omitted from these specifications or from the official drawings, but which is necessary for the proper completion of the vessel, is to be supplied by the contractor without extra charge.

o

www.ingramcontent.com/pod-product-compliance
Lightning Source LLC
Chambersburg PA
CBHW032132160426
43197CB00008B/613